HERRICK DISTRICT LIBRARY
Holland, MI

MAY 2 4 2024

US MILITARY ENCYCLOPEDIAS

THE MILITARY WEAPONS ENCYCLOPEDIA

BY ARNOLD RINGSTAD

Encyclopedias

An Imprint of Abdo Reference
abdobooks.com

TABLE OF CONTENTS

WEAPONS OF THE US MILITARY 4
BAYONETS
M9 Bayonet 6
OKC-3S Bayonet 8
HANDGUNS
M9 10
M17 12
ASSAULT RIFLES
FN SCAR-L/H 14
M14 Enhanced Battle Rifle 16
M16A4 18
M27 Infantry Automatic Rifle 20
XM7 22
CARBINES
M4 Carbine 24
SNIPER RIFLES
M24 Sniper Weapon System 26
M107 28
M110 Semi-Automatic Sniper System ... 30
M2010 Enhanced Sniper Rifle 32
SHOTGUNS
M26 Modular Accessory Shotgun System 34
M500 36
M1014 38
SUBMACHINE GUNS
MP5 40
MP7 42
MACHINE GUNS
M2 Browning 44
M240 46
M249 48
XM250 Light Machine Gun 50

GRENADES
M18 Smoke Grenade 52
M67 Fragmentation Grenade 54
M83 Smoke Grenade 56
M84 Flash Bang Grenade 58
GRENADE LAUNCHERS
M203 60
M320 62
Mk-19 Grenade Machine Gun 64
ROCKET LAUNCHERS
BGM-71 TOW 66
FGM-148 Javelin 68
FIM-92 Stinger 70
M3 MAAWS 72
M136 AT4 74
Mk-153 SMAW 76
MINES
M18A1 Claymore 78
MORTARS
M120 80
M224 82
M252 84
HOWITZERS
M109 86
M119 88
M777 90
ROCKET ARTILLERY
M142 HIMARS 92
M270 MLRS 94
BOMBS
AGM-154 Joint Standoff Weapon 96
GBU-15 98
GBU-39 Small Diameter Bomb 100

GBU-43/B MOAB..102
GBU-57 Massive
    Ordnance Penetrator 104
JDAM 106
Paveway 108

AIRCRAFT CANNONS
GAU-8 Avenger 110
M61 Vulcan..112
M134 Minigun..114
M230 Chain Gun......................................116

CRUISE MISSILES
AGM-86 ALCM...118
AGM-129 Advanced Cruise Missile......120
AGM-158 JASSM.......................................122
BGM-109 Tomahawk...............................124

AIR-TO-AIR MISSILES
AIM-7 Sparrow 126
AIM-9 Sidewinder 128
AIM-120 AMRAAM...................................130

SURFACE-TO-AIR MISSILES
RIM-116 Rolling Airframe Missile 132
RIM-162 Evolved
    Sea Sparrow Missile.............................134

AIR-TO-SURFACE MISSILES
Advanced Precision
    Kill Weapon System 136
AGM-65 Maverick.....................................138
AGM-84K SLAM-ER..................................140
AGM-88 HARM 142
AGM-114 Hellfire 144
AGM-176 Griffin..146

ANTI-SHIP MISSILES
AGM-119 Penguin 148
AGM-158C LRASM....................................150

RGM/AGM/UGM-84 Harpoon................152

TORPEDOES
Mk-48 154
Mk-50 156
Mk-54 158

NAVAL MINES
Quickstrike Mine 160
Submarine-Launched Mobile Mine 162

NAVAL GUNS
Mk-38 164
Mk-45 166

BALLISTIC MISSILES
LGM-30 Minuteman III 168
UGM-133 Trident II 170

NUCLEAR WEAPONS
B61 Nuclear Bomb....................................172
B83 Nuclear Bomb....................................174

MISSILE DEFENSE
Aegis Ballistic Missile Defense................176
Ground-based Midcourse Defense 178
MIM-104 Patriot..180
Mk-15 Phalanx CIWS................................182
NASAMS 184
THAAD 186

GLOSSARY 188
TO LEARN MORE.............................. 189
INDEX 190
PHOTO CREDITS 191

WEAPONS OF THE US MILITARY

The US military has a massive arsenal of weapons. These tools let the military fight and win battles on land, in the air, and at sea. Powerful weapons alone aren't enough, though. Soldiers, Marines, sailors, pilots, and other members of the military are highly trained. They use their knowledge of these weapons to defeat enemies and complete missions.

TYPES OF WEAPONS

On land, troops use bayonets, handguns, and shotguns for close-up fighting. Assault rifles and machine guns allow soldiers to attack enemy forces at longer range. Troops use sniper rifles to hit enemies at extreme distances. Mines, grenade launchers, and rocket launchers make it possible for individual soldiers to destroy enemy vehicles. Howitzers and rockets can take out enemy buildings from many miles away.

In the air, fighter jets destroy other planes with air-to-air missiles. Bombers and attack planes hit ground targets using bombs and air-to-surface missiles. Cannons and chain guns give planes and helicopters rapid-firing weapons to use against enemies on the ground.

At sea, ships fire powerful naval guns and launch anti-ship missiles. They also fire cruise missiles to hit distant targets on land. Beneath the waves, submarines shoot torpedoes at enemy vessels and place underwater mines. They also launch ballistic missiles that can travel thousands of miles.

Nuclear weapons are the most powerful weapons in the world. They have only been used twice in combat. The United States dropped two nuclear weapons on Japan just before the end of World War II (1939–1945). Today, nations hope that building up a nuclear arsenal will prevent wars that could lead to devastating nuclear attacks.

Some weapons are used for defense. At land and at sea, these systems track incoming enemy missiles with radar. Then they use missiles or rapid-firing guns to shoot down these threats.

Whether a soldier is firing a pistol or a ballistic missile, training and practice help ensure the weapon is being used correctly. Soldiers learn how to use these weapons individually and in teams to complete missions and win battles. A combination of impressive technology and intense preparation helps make the US military a deadly fighting force.

BAYONETS

M9 BAYONET

The M9 bayonet is a bladed weapon used by the US Army. Soldiers can attach it to M16 rifles or M4 carbines to use it as a bayonet. They can also use it as a handheld weapon by itself.

The M9 is useful for more than just fighting. It also works as a utility knife or a small saw, slicing, chopping, or cutting things on the battlefield. The M9 can even be used as a wire cutter. The knife attaches to its sheath to form a scissor-like shape. Soldiers can then use it to slice through barbed wire or fencing.

HISTORY

In the mid-1980s, the US Army held a competition to replace the M7 bayonet. A company called Phrobis beat five other companies in the final selection. Its new bayonet became known as the M9. It entered service in 1987.

It wasn't long before the M9 saw action in the field. Soldiers used it in 1989 in Operation Just Cause in Panama. A group of Army Rangers used the wire-cutting feature to slice through a chain-link fence and complete a mission.

The hole in the M9's blade lets it connect to the sheath so it can be used for cutting on the battlefield.

Attaching a bayonet to a rifle gives soldiers the flexibility to fight close up or at long range.

BAYONETS

Marines train for intense hand-to-hand fighting.

OKC-3S BAYONET

The OKC-3S bayonet is used by the US Marine Corps. The weapon is also known as the multipurpose bayonet. It attaches to the M16 and M4 rifles that Marines carry. It is designed to function well in both extreme cold and intense heat. The bayonet has an 8-inch (20 cm) blade made of tough carbon steel. In addition to being deadly in combat, the bayonet is also useful for cutting or chopping on the battlefield.

ORIGINS

The OKC-3S is made by the Ontario Knife Company of New York. The company's name is where the *OKC* in the bayonet's name comes from. The weapon replaced the earlier M7 as the main bayonet used by the Marines. When searching for its new bayonet, the Marine Corps tested a total of 33 models. The OKC-3S received high ratings in almost every category.

The OKC-3S entered service with the Marines in 2003.

HANDGUNS

M9

The M9 is a 9-mm pistol. Made by the Italian company Beretta, for many years it was the standard gun used by all branches of the US military. The M9 is a semiautomatic pistol. Its magazine holds 15 rounds.

Sailors practice firing the M9 at sea.

Soldiers train on how to quickly and safely reload their pistols.

In 1978, the US military began searching for a handgun to replace the M1911, which dated to the year 1911. In 1985, the Army selected Beretta's 92 SB-F as the winner. It gave the pistol the official name M9.

IMPROVING THE M9

The M9 has seen many improvements over the years. The M9A1 was introduced in 2006. It added a rail below the barrel where soldiers could attach accessories, such as flashlights or lasers. Beretta also added a new coating that made the gun more durable. This made the M9 more reliable in desert environments, such as those found in Iraq and Afghanistan. Another new version, the M9A3, arrived in 2015. It added more places to attach accessories and featured a larger magazine holding 17 rounds.

HANDGUNS

M17

In 2008, the Air Force and Army began searching for a pistol to replace the M9. The M9 dated back to the 1980s, and the military wanted a newer pistol with better accuracy and reliability. An official competition began in 2011. The company SIG Sauer entered its P320 pistol. In 2017, it was announced as

Accessories under the barrel of the M17 add more functions to the pistol.

SIG Sauer won a contract worth $580 million for producing the M17, along with accessories and ammunition.

the winner. The gun entered service as the M17. The military also ordered a smaller version called the M18.

The M17 is a 9-mm semiautomatic pistol. It has a magazine capacity of 17 rounds. An accessory rail lets users add flashlights or lasers. Compared with the M9, the M17 is lighter. This makes the gun easier to handle, as does its improved grip design.

A RUGGED GUN

During testing, the Army studied how well the pistol held up under harsh conditions. Testers dropped the gun onto the ground from different heights. They exposed it to salt water, which can easily damage firearms. Through dust, sand, heat, and cold, the M17 remained reliable.

ASSAULT RIFLES

FN SCAR-L/H

The US Special Operations Command (SOCOM) issued a request for a new rifle in 2003. The Belgian company FN Herstal developed the SCAR assault rifle, and it won the competition in 2004. The gun eventually entered service with special forces troops in 2009.

The SCAR is gas-operated. This means that the gas released when firing a round provides the energy to load the next round. The gun is a selective-fire weapon, so the user can decide to fire semiautomatic or fully automatic. In the gun's

A suppressor on the end of the FN SCAR-H reduces the noise from firing.

US Army Green Berets trained on SCAR rifles in 2023 in Bosnia and Herzegovina.

semiautomatic mode, it fires once each time the trigger is pulled. In fully automatic, the gun keeps firing as long as the trigger is held.

LIGHT AND HEAVY

There are two versions of the rifle: the SCAR-L (light) and SCAR-H (heavy). The L model fires 5.56-mm rounds, and the H model uses 7.62-mm rounds. Because the L uses smaller ammunition, it can hold more rounds. The SCAR-L uses 30-round magazines, while the SCAR-H holds 20-round magazines.

ASSAULT RIFLES

M14 ENHANCED BATTLE RIFLE

The original M14 rifle entered service with the US military in the late 1950s. Around the year 2000, the gun received a significant upgrade. Engineers created a lighter, smaller version of the rifle for use by US Navy SEALs and other soldiers. The gun had a shorter barrel, a better grip, and rails for adding accessories. It became known as the M14 Enhanced Battle Rifle (EBR).

LONG-RANGE FIGHTING

The M14 EBR is designed for longer-range shooting than the M16 rifle and the M4 carbine. The gas-operated weapon fires

A sailor practices firing an M14 EBR from the edge of an aircraft carrier.

Soldiers use huge firing ranges to train on long-range shooting with the M14 EBR.

7.62-mm rounds. It holds a 20-round magazine. An attached scope helps improve accuracy at long range. The gun also has a folding bipod in the front. This helps the soldier keep the rifle steady for challenging shots. The M14 EBR can hit targets at a range of 2,625 feet (800 m), but it is also good for close-range combat if needed.

ASSAULT RIFLES

M16A4

The M16A4 is a gas-operated rifle. It fires 5.56-mm rounds and uses 30-round magazines. The gun features a rail for adding scopes or night vision equipment. It can reliably hit targets at a range of 1,970 feet (600 m). Soldiers can attach a bayonet to the front of the M16A4. They can also install an M203 grenade launcher under the barrel.

HISTORY

The M16 rifle was introduced during the Vietnam War (1954–1975). Early models faced reliability problems. But the gun saw many upgrades over the following decades. The latest model is the

Soldiers train to fire the M16A4 in all kinds of conditions, including while wearing gas masks.

The US military uses the M16A4 for recruit training.

M16A4, introduced in 1998. It was the standard Marine rifle from then until 2015, when it was replaced by the M4 carbine. The M16A4 is still used by some troops.

The original M16 had fully automatic fire. But recoil meant that accuracy dropped sharply after a few shots. The M16A4 can fire semiautomatically or in three-round bursts. This provides more accuracy and wastes less ammunition.

ASSAULT RIFLES

Soldiers can steady the M27 on a nearby surface rather than using the attached bipod.

M27 INFANTRY AUTOMATIC RIFLE

The German company Heckler & Koch created the HK416 assault rifle in the 1990s. It was an upgraded version of the M4 carbine. In 2010, the company made a new model called the HK416D. The next year, the US Marine Corps began using the gun under the name M27.

The M27 is a gas-operated automatic rifle. It fires 5.56-mm rounds. The M27 has selective fire, so troops can fire it in a semiautomatic or fully automatic mode. The Marines used it to replace some M249 light machine guns. Each four-person squad has one M27 rifle.

PROS AND CONS

The M27 uses 30-round magazines. It received some criticism because the M249 uses much larger 200-round belts of ammunition. This means the M27 needs to reload more often. However, the gun also has much better reliability and accuracy than the M249. It comes standard with a scope that provides 3.5x magnification, meaning it makes objects look 3.5 times larger. It also features multiple rails for adding accessories.

The scope of an M27 gives Marines a better view of distant targets.

ASSAULT RIFLES

XM7

In 2022, the US Army made a major announcement. It had chosen a new weapon to replace the M4 carbine. This new rifle is called the XM7. It is made by SIG Sauer.

The XM7 uses 6.8-mm ammunition. These rounds are more accurate and deadlier than the 5.56-mm ammunition used in the M4. The XM7 holds a 20-round magazine.

The XM7 comes with a suppressor as a standard feature.

A HIGH-TECH SCOPE

One of the XM7's most advanced features is its scope. It is called the XM157 fire control system. As with any other scope, soldiers can simply use it to aim as they always have. But it also includes impressive new technology. The XM157 contains sensors, a computer, and a digital display. Soldiers can accurately calculate where their shots will hit.

Soldiers are even able to share data with each other. For example, a soldier could look at a target and mark it using the scope. Other soldiers could then look through their scopes and see what has been marked.

If the XM7's advanced scope runs out of battery, soldiers can still use it as a normal scope.

CARBINES

M4 CARBINE

Carbines are essentially short rifles. Their small size makes them easier to handle than larger guns. The M4 carbine is one example of this. It is a shortened version of the M16A2 rifle. The US Army began using the M4 in 1994. Since then, it has become a common weapon across the US military.

The M4 is made by the Colt company. Selective fire allows soldiers to shoot semiautomatically or in three-round bursts. Like the M16, the M4 fires 5.56-mm rounds and uses 30-round magazines. However, the

The M4 is shorter than full-sized rifles, making it easier to use in tight spaces.

The automatic mode of the M4A1 gives the gun more firepower, which is useful when setting up a defensive position.

shorter barrel means the M4 has less range. The carbine's effective range is about 1,180 feet (360 m).

DIFFERENT VERSIONS

A few different versions of the M4 exist. The M4A1 is a carbine version of the M16A3. Like the M16A3, it has a fully automatic firing mode instead of using three-round bursts. Another model is the M4 Commando. It has an even shorter barrel than the standard M4. Some special forces troops use this weapon.

SNIPER RIFLES

Snipers use camouflage to help them sneak closer to enemy targets.

M24 SNIPER WEAPON SYSTEM

The M24 Sniper Weapon System entered service with the US Army in 1988. It is based on the Remington Model 700, a civilian hunting rifle. The gun became the first Army weapon designed specifically for snipers.

The rifle fires 7.62-mm rounds to a range of about 2,625 feet (800 m). Its magazines hold either five or ten rounds. Like other sniper rifles, the gun includes a scope for long-range shots. It also has iron sights as a backup.

BOLT ACTION

The M24 is a bolt-action rifle. This means it must be reloaded after each shot. Users pull back on a small handle to eject the used casing from the round they just fired. Then they push forward to load the next round.

The rifle has seen upgrades over the years. The M24A2 model includes rails for adding accessories to the gun. The M24A3 uses larger ammunition and is meant for longer-range shooting.

Developing sniping skills takes a lot of practice.

SNIPER RIFLES

M107

In 1990, the US Navy and Marine Corps started using a powerful sniper rifle called the M82A1. The Army was impressed and decided to order a large rifle of its own. After testing weapons, it ended up using the M82A1 with minor changes. The Army renamed the weapon the M107, and it began production in 2002. An updated version, the M107A1, arrived in 2010. It featured reduced weight and other improvements.

A fully loaded M107 weighs about 35 pounds (16 kg).

The M107's power and accuracy let soldiers hit enemies at very long ranges.

A POWERFUL RIFLE

The M107 fires .50-caliber ammunition. These rounds have extreme power and range. They can hit targets at 1.24 miles (2 km). The M107 is deadly against enemy soldiers. The gun's long range makes it useful against snipers. Soldiers can use it to hit enemies who lack the range to fire back.

The gun is also effective against lightly armored vehicles, aircraft on the ground, and other military equipment. The weapon's power means it is very loud. Soldiers must wear ear protection when shooting it.

SNIPER RIFLES

M110 SEMI-AUTOMATIC SNIPER SYSTEM

The M110 Semi-Automatic Sniper System is a sniper rifle using 7.62-mm rounds. The gun can use a magazine holding either 10 or 20 rounds. It has a maximum range of about 2,625 feet (800 m). The rifle includes a suppressor, making it harder for enemies to find the sniper who has taken a shot.

The rifle has a scope that ranges from 3.5x to 10x magnification. This helps a soldier aim at both near and far targets. It also has a bipod, helping the shooter steady the gun for better accuracy.

THE SNIPER TEAM

The M110 comes with a separate piece of equipment called an M151 spotting scope. Snipers usually work with a second person

The M110 has a safety switch on both sides, making it easy for both right-handed and left-handed soldiers to use.

The M151 spotting scope is made by the company Leupold.

called a spotter. The sniper and the spotter work together to get into position. The spotter studies the target with the spotting scope. This person helps the sniper plan a shot. He or she also helps the sniper adjust if the first shot misses.

SNIPER RIFLES

M2010 ENHANCED SNIPER RIFLE

During the Afghanistan War (2001–2021), soldiers faced battlefields with wide-open spaces. They wanted sniper rifles with better range. The military decided to upgrade the M24 instead of creating an entirely new weapon. The result was the M2010 Enhanced Sniper Rifle. The *2010* in its name represents the planned year the gun would enter service, but it ended up arriving in 2011.

UPGRADES AND FEATURES

The M2010 features many upgrades over the M24. One of the biggest is its new ammunition. The M2010 replaces the old rifle's 7.62-mm

When used in snowy environments, rifles may be painted white to make them harder for enemies to spot.

Nighttime training gives soldiers a chance to practice using the M2010's night vision scope.

rounds with more powerful .300 Winchester Magnum ammunition. The M2010's range is 3,940 feet (1,200 m), 50 percent farther than the original M24.

Like the M24, the M2010 is a bolt-action rifle. The gun includes a suppressor. It also has an advanced scope with night vision. The rifle's rail system lets soldiers add more accessories.

SHOTGUNS

M26 MODULAR ACCESSORY SHOTGUN SYSTEM

The US Army began seeking an accessory shotgun in the late 1990s. This type of weapon is attached to another gun. The result of the Army's search was the M26 Modular Accessory Shotgun System (MASS). It entered service in 2008.

The M26 is mounted under the barrel of an M4 carbine. It is a bolt-action 12-gauge shotgun with a small magazine holding three or five rounds. The bolt can be mounted on either side, making it easy to handle for both right-handed and left-handed soldiers. The M26 also comes in a stand-alone version with its own grip and stock.

Attaching the M26 below a soldier's main rifle provides flexibility in combat.

A stand-alone version of the M26 gives the weapon its own stock and grip.

USING THE M26

The M26 gives soldiers the benefits of a shotgun without needing to carry an extra gun. Shotguns can be used to break down doors. To be used for crowd control, they can be loaded with rounds designed not to kill people. They can also be used for fighting at very close range. Soldiers can use the M26 without having to put away their main weapons.

SHOTGUNS

M500

The M500 is a pump-action shotgun. This means the soldier slides the handguard back to eject the spent cartridge. Sliding it forward again loads the next cartridge. The M500 holds a five-round magazine. Different types of ammunition are used for short-range combat and breaching doors. Rounds designed not to kill people can be used to control crowds.

HISTORY AND UPGRADES

The M500 is made by the firearm company Mossberg. It has been in service with the Marines since the 1980s. The gun has

Soldiers in Afghanistan held a shooting competition with the M500 in 2014.

Some types of shotgun shells blast a wide pattern of small, deadly projectiles.

seen various upgrades over the years. A more durable version that replaced some plastic parts with metal ones was called the M590.

Around 2009, the Army added the Military Enhancement Kit (MEK) to the M500. The MEK includes a new barrel, more options for stocks, and places to attach accessories. Older models of the shotgun had no simple way to attach things such as flashlights, lasers, and forward grips.

Practicing on firing ranges helps Marines maintain their marksmanship skills.

M1014

In the late 1990s, the US military began looking for a new shotgun. The Marine Corps put out a list of requirements for the new gun. The Italian shotgun company Benelli entered its M4 shotgun in the competition. The Marines chose Benelli's gun, and it entered service in 1999 as the M1014.

SEMIAUTOMATIC SHOTGUN

Earlier military shotguns were pump-action weapons. The M1014 is semiautomatic, so it fires and reloads itself with each pull of the trigger. Some semiautomatic shotguns are delicate, but the M1014 is designed to be rugged. The reloading mechanism can stand up to sand and dust.

However, this mechanism requires powerful combat shells to work. This means the M1014 cannot be used with less-lethal ammunition. These shells do not release enough energy to reload the gun. As a result, the M1014 is used only for combat. It is a devastating weapon at close range. Troops can fire it repeatedly to quickly clear a room.

Troops train on the process of breaching doors with an M1014 shotgun.

SUBMACHINE GUNS

Members of the security teams for top military officials train with the MP5.

MP5

The MP5 is a submachine gun. This type of gun is a lightweight automatic weapon. It fires the small rounds used in pistols, rather than the large rounds used in full-size machine guns. Submachine guns provide a lot of firepower in a small package.

The MP5 was developed by a German company called Heckler & Koch in the 1960s. The gun fires 9-mm rounds. Selective-fire modes let users discharge single shots, three-round bursts, or fully automatically. In automatic mode, the MP5 shoots at a rate of 800 rounds per minute.

SPECIAL FORCES

This weapon became popular with special forces troops across the globe, including the US Navy SEALs. Troops like it for its small size, durability, and low recoil. It is easy to aim in tight spaces and keep shots on target.

Specialized models of the MP5 have additional features. One version has a built-in suppressor to reduce noise. Another is even more compact than the normal MP5. It is small enough to carry in a briefcase.

A folding stock helps make the MP5 even smaller and easier to carry.

SUBMACHINE GUNS

MP7

Like the MP5, the MP7 submachine gun is made by Heckler & Koch. In the 1980s, militaries around the world were looking for a compact weapon that would be effective against body armor. Heckler & Koch developed the MP7 to meet this goal.

The MP7 is like a cross between a submachine gun and a carbine. It fires 4.6-mm rounds similar in size to those used in pistols. But the rounds are designed to penetrate targets more like those used in rifles. The small rounds mean that troops can carry more ammunition. A 40-round MP7 magazine is about the same size as a 30-round MP5 magazine. The light rounds also reduce recoil, improving accuracy.

US Marines trained with the MP7 during an exercise in Romania in 2016.

The MP7 is especially effective for close-range combat.

A FAMOUS RAID

The gun is in service with special forces, including the US Navy SEALs. In 2011, a team of SEALs found and killed terrorist leader Osama bin Laden in Pakistan. Some of the SEALs reportedly used MP7s during the famous mission.

MACHINE GUNS

M2 BROWNING

The M2 Browning is a heavy machine gun firing .50-caliber rounds. In automatic mode, it can shoot up to 850 rounds a minute. Its high-powered rounds travel a long way. The gun has

The size and weight of the M2 make it a good vehicle-mounted weapon.

The M2's history with the US military goes back nearly a century.

a maximum range of about 4.2 miles (6.8 km). It is a belt-fed machine gun. This means it uses a continuous belt of rounds rather than a magazine. The M2 is used by the Army, Navy, Air Force, and Marine Corps.

The Browning is used for suppressive fire. Soldiers shoot it to force enemies to take cover. This allows friendly soldiers to advance. The gun can be used on its own, or it can be mounted to vehicles. It is powerful enough to damage vehicles, including aircraft and boats.

HISTORY

The M2 Browning was originally developed by American firearm designer John Browning in the 1920s. It has been used in major US conflicts since then. In 2011, the military began using the upgraded M2A1. Among other features, the new version added a flash hider. This reduces the gun's muzzle flash, making it harder for enemies to spot the gun in the dark.

MACHINE GUNS

M240

The M240 machine gun is a US version of the FN MAG. This Belgian machine gun dates back to 1957. The US Army adopted it as the M240 in 1977. It has also been used by the Navy, Marine Corps, and Coast Guard.

The M240 fires 7.62-mm rounds. It is a belt-fed weapon, loaded with linked belts of rounds up to 250 rounds long. The gun has three selective fire modes. These modes allow the weapon to fire at 750, 850, or 950 rounds per minute.

DIFFERENT VERSIONS

The M240 has multiple versions. The M240B includes

The M240B can be held in a soldier's hands or mounted on vehicles.

> Soldiers practice using the M240L for suppressing fire at a training range in Hawaii.

improvements that reduce the gun's recoil. The M240C is a version mounted on vehicles such as the M2 Bradley infantry fighting vehicle and the M1 Abrams tank. The M240L is a lighter version. By using strong, lightweight titanium, it weighs almost 20 percent less than the M240B.

MACHINE GUNS

M249

The M249 is a light machine gun. It fires 5.56-mm rounds at a rate of up to 1,000 rounds per minute. The gun can be held in the hands while firing, or it can be mounted on a vehicle. A folding bipod on the front helps soldiers keep the gun stable for more accurate fire.

The weapon uses 200-round magazines. However, because it fires the same rounds as the M16, it can also use standard 30-round M16 magazines if needed. The heavy gun has a handle on top to make it easier to carry into battle. The M249's barrel heats up while firing and sometimes needs to

The M249 can be fed by belts of ammunition rather than magazines.

Soldiers work together to quickly swap barrels and keep firing.

be swapped. It takes a trained soldier about seven seconds to replace a barrel.

HISTORY

The M249 is a US version of the Belgian FN MINIMI machine gun. It entered service with the US Army in 1982. The light machine gun has been used in combat in Iraq, Afghanistan, and elsewhere.

MACHINE GUNS

The XM250's bipod helps give soldiers steadier aim when firing.

XM250 LIGHT MACHINE GUN

The XM250 light machine gun emerged from the same program as the XM7 assault rifle. Like the XM7, it is made by SIG Sauer. The XM250 uses the same new 6.8-mm ammunition as the XM7. The military planned to replace the M249 with this new machine gun starting in late 2023.

The XM250 went through more than two years of testing. More than 1,000 soldiers tried the weapon and gave feedback in that time. The Army planned to order as many as 13,000 XM250 light machine guns by the early 2030s. It planned to keep the M249 in service until it could be completely replaced by the XM250.

50

FEATURES

The new weapon is about 4 pounds (1.8 kg) lighter than the M249. It also has better accuracy and is easier to handle. The gun features a built-in suppressor to reduce noise. This makes it easier for soldiers in a squad to hear each other during a fight. The XM250 uses 100-round pouches of ammunition.

SIG Sauer showed off the XM250 at a European military trade show in 2022.

GRENADES

M18 SMOKE GRENADE

Smoke grenades give off thick clouds of smoke when thrown. They have several military uses. They can signal other troops on the ground. They can also signal airplanes and helicopters flying above. The smoke can mark a target for friendly forces to attack. It can also be used to mark a landing zone for incoming helicopters. Finally, smoke grenades can hide friendly forces from view so they can safely move around the battlefield. The M18 smoke grenade is used for all of these tasks.

    The M18 weighs about 19 ounces (539 g). A typical soldier can throw the cylindrical grenade about 100 feet (30 m). Soldiers must be careful where they throw the grenade, because the heat it releases may start a fire in dry conditions.

Soldiers toss the M18 smoke grenade by hand to mark targets or obscure the battlefield.

COLOR OPTIONS

Multiple color options are available. M18 grenades can produce violet, yellow, green, or red smoke. When used for signaling, soldiers may plan ahead for these colors to have different meanings. A grenade gives off its colored smoke for about 90 seconds.

Colored smoke has multiple uses in battle.

GRENADES

Practice versions of the M67, known as the M69, let soldiers safely train on throwing these grenades.

M67 FRAGMENTATION GRENADE

The M67 fragmentation grenade is used in close-range combat. The grenade is a steel sphere 2.5 inches (6.4 cm) wide. It is filled with an explosive called Composition B. The user pulls a pin and then tosses the grenade. About four or five seconds later, it explodes.

When the explosive goes off, it sends pieces of the steel sphere flying in all directions. These fragments usually kill people within 16 feet (5 m). They injure people within 49 feet (15 m). However, chunks of metal can be launched as far as 755 feet (230 m) away. A typical soldier can throw the M67 about 130 feet (40 m). This means it is important to take cover after throwing.

HISTORY

The M67 was developed in the 1960s. It replaced the earlier M26 grenade. It was designed to be smaller and lighter. This makes it easier to throw a long distance. The M67 is shaped similarly to a baseball, so many soldiers are able to quickly learn how to throw the grenade.

Soldiers must be careful not to get caught in the blast after tossing a grenade.

GRENADES

M83 SMOKE GRENADE

The M83 smoke grenade produces thick white smoke. The smoke can hide friendly forces from enemy view. It can also be used to signal fellow soldiers. The grenade is made of metal and weighs about 16 ounces (454 g).

Earlier types of smoke grenades used a chemical called hexachloroethane to create smoke. But it was discovered that this chemical can have severe health effects. Vomiting, organ damage, and even death can occur. Scientists developed new smoke grenades using terephthalic acid instead. This makes the grenade safer for the troops using it.

SETTING IT OFF

To create smoke, the soldier pulls the grenade's ring, then tosses the grenade. The fuse burns at a high temperature. Within a few seconds, smoke begins coming out. The M83 keeps producing white smoke for up to 90 seconds. The M83 and the M18 both use the M201A1 fuse to set off this process.

The M83 has a cylindrical shape.

Smoke can be used to conceal friendly troops.

GRENADES

M84 FLASH BANG GRENADE

Flash bang grenades are not designed to kill people. When thrown, they create an extremely loud noise and a bright flash. The noise can be as loud as 180 decibels. This is louder than standing right next to a jet airplane taking off. The effects of the flash bang prevent enemy troops from seeing or hearing anything for a few seconds. This can provide a big advantage in a fight.

The M84 can instantly disorient everyone in a room.

A Marine prepares to toss a practice version of the M84 into a room during a training exercise.

The M84 flash bang grenade has a steel body filled with the chemicals that create the flash bang effect. A fuse on top sets off the grenade. Venting holes along the sides allow the light and noise to escape. These grenades can cause damage to hearing and vision.

USING THE M84

Soldiers use the M84 to clear out rooms. If enemies are in a room, a soldier might toss one of these grenades through the doorway. The grenade goes off, and the room is filled with light and sound. The enemy troops inside are stunned. The friendly troops can move in and clear out the enemies.

GRENADE LAUNCHERS

M203

Grenade launchers have a long history in the US military. They allow troops to fire grenades much farther than they can be thrown. Some of these grenade launchers attach to a soldier's rifle. During World War II and the Korean War (1950–1953), US troops used the M7 launcher attached to the M1 Garand rifle. It was an effective weapon, but the rifle could not fire normally while it was attached.

The M203 adds explosive firepower to a soldier's main weapon.

In the late 1960s, the new M203 grenade launcher entered service during the Vietnam War. It attached to the M16

assault rifle below the barrel. This meant the gun could still be fired normally.

FIRING THE M203

The M203 is a single-shot grenade launcher. It must be loaded each time it is fired. It shoots 40-mm grenade rounds. These can be lethal fragmentation grenades. They may also carry smoke or tear gas. The weapon has a range of 1,150 feet (350 m). A later version, the M203A1, works with the M4 carbine.

Blue-colored practice rounds are used to safely train on firing the M203.

GRENADE LAUNCHERS

M320

In 2004, the Army put out a call for a new grenade launcher. This new weapon would replace the existing M203. The German company Heckler & Koch submitted its AG36 grenade launcher. The military selected it, and the weapon entered service in 2009 as the M320.

The M320 is used by the Army and the Marine Corps. It is a single-shot grenade launcher that is loaded from the side. It fires the same 40-mm grenades designed for the M203. Soldiers can accurately hit targets about 490 feet (150 m) away using these explosive rounds.

The M320 is able to fire newer, more advanced rounds that don't work with the M203.

A sight helps soldiers fire the M320 at the correct angle to hit the target.

FIRING OPTIONS

Like the M203, the M320 can be attached to a soldier's main weapon. This includes the M16 or M27 assault rifles or the M4 carbine. The grenade launcher can also be used in a stand-alone mode. It has a folding grip, a detachable stock, and a sight. These features let soldiers fire the grenade launcher on its own.

GRENADE LAUNCHERS

Linked belts of grenades let the Mk-19 keep up a rapid rate of fire.

MK-19 GRENADE MACHINE GUN

The Mark-19, or Mk-19, grenade machine gun is an automatic grenade launcher. The weapon fires 40-mm grenades to a maximum range of about 1.2 miles (2 km). The grenades are fed into the Mk-19 in belts of 32 or 48 rounds.

Soldiers hold grips at the back of the weapon and use a sight to aim and fire. When the grenades explode, they are powerful enough to kill within 16 feet (5 m) and wound within 49 feet (15 m). The Mk-19 allows troops to rapidly fire dozens of these grenades at the enemy.

DEVELOPMENT AND USE

This grenade launcher was developed in the 1960s and saw use in the Vietnam War. It has had several upgrades over the years. Today, it is in service with the Air Force, Marine Corps, Army, and Navy. Soldiers can fire it from a tripod on the ground. They can also mount it on vehicles. US troops can use the Mk-19 to take down helicopters and lightly armored vehicles.

The Mk-19 enables troops to quickly fire a tremendous amount of explosive power.

ROCKET LAUNCHERS

An advanced optical sight helps operators steer TOW missiles to their targets.

BGM-71 TOW

The BGM-71 is a rocket launcher used against enemy tanks and structures. It is known as TOW because it fires Tube-launched, Optically tracked, Wire-guided missiles. The weapon is used by the Army and Marine Corps.

The TOW fires missiles out of a tube. The launcher can be mounted on a tripod on the ground. It can also be mounted to trucks, armored vehicles, and even helicopters. Launchers on vehicles may have up to four tubes for launching multiple missiles.

TRACK AND FIRE

The operator tracks targets optically, meaning by sight. After locating a target in the weapon's sight, the operator fires the missile. A solid rocket motor propels the missile out of the tube. A long, thin wire connects the missile to the launcher. It unspools as the rocket flies. The wire carries signals from the launcher to the missile. It allows the operator to guide the missile by keeping the enemy target in the sight. Newer versions also offer a wireless guidance mode.

US Army troops fire a TOW missile from a Humvee-mounted launcher.

ROCKET LAUNCHERS

The Javelin missile is propelled out of the launch tube before its rocket motor ignites, improving safety for the operator.

FGM-148 JAVELIN

The FGM-148 Javelin is a portable anti-tank rocket launcher. The Army and the Marine Corps use it. The Javelin entered service in 1996 and has been a key weapon for US forces since then.

The Javelin is made up of two parts. The first is the Command Launch Unit (CLU). The second is the launch tube assembly. The CLU is used to find targets, lock onto them, and fire. It is reusable. The launch tube assembly is a sealed tube containing one missile. It is disposable and must be replaced for each shot.

FIRING THE JAVELIN

This powerful rocket launcher has two attack modes. It can fire directly at targets if needed. But it is more commonly used in top attack mode. After launching, the missile shoots high into the air. Then it attacks the target from above. This allows it to hit a tank's armor on top, where it is thinnest. After launching, the missile guides itself to the target. This means the soldier who fired it can quickly move to a new position in case the enemy starts firing back.

The advanced targeting system in the CLU locks onto enemy targets.

ROCKET LAUNCHERS

FIM-92 STINGER

The FIM-92 Stinger is a shoulder-fired missile launcher used against aircraft. It is a type of weapon known as a man-portable air defense system (MANPADS). It can attack low-flying airplanes and helicopters at a range of up to 5 miles (8 km). The Stinger is used by the Army and Marine Corps.

The system includes a launch tube, a cooling unit, a sight, and an Identification Friend or Foe (IFF) device. The launcher uses a heat-seeking sensor to target aircraft. The cooling unit is inserted just before firing to chill this sensor. The IFF device helps the operator determine

The Stinger missile helps ground troops defend themselves from enemy aircraft.

Virtual reality training allows soldiers to practice firing the Stinger without having to waste the expensive missiles.

whether an aircraft is friendly or hostile. After firing, the missile uses fins to steer itself to the target.

HISTORY AND UPGRADES

The Stinger was developed in the late 1970s and entered service in 1981. It has seen many upgrades over the years. These changes helped the missile resist flares, which enemy aircraft release to distract infrared sensors. Upgrades also made the Stinger better at hitting small targets, including drones.

ROCKET LAUNCHERS

Soldiers firing the M3 must be sure to stand clear of the backblast.

M3 MAAWS

The M3 MAAWS is a portable rocket launcher. Its name stands for Multipurpose Anti-armor Anti-personnel Weapon System. The M3 is in service with the Army and the Marine Corps. It is a durable weapon, able to survive exposure to salt water and the forces of parachuting from an airplane.

This weapon usually has a crew of two. One soldier carries and fires the launcher. The other is responsible for carrying extra ammunition and reloading the M3. Working together, they can attack targets at ranges of up to 2,625 feet (800 m).

MANY USES

As its name suggests, the M3 MAAWS has many uses. It can launch a variety of 84-mm rounds. Some of its rounds create smoke to give cover to friendly troops. Others are bright flares that burst in midair, lighting up a dark battlefield. Anti-tank rounds are designed to destroy armored vehicles, and anti-personnel rounds are meant to hit enemy troops.

The M3 MAAWS was originally developed in Sweden.

ROCKET LAUNCHERS

M136 AT4

The M136 is a portable anti-armor rocket launcher used by the Army. The rocket and launcher come together in a self-contained package. Soldiers fire the weapon from the right shoulder. After firing once, the disposable launcher is discarded.

The rockets have fins that pop out in midair to keep the rocket stable. However, the rocket does not steer in midair. It simply goes directly where the soldier aimed it. Each rocket contains 15.5 ounces (439 g) of explosive in a shaped charge. When hitting a target, this charge focuses the explosion into a narrow jet of gas that can punch through armor. The M136 can penetrate armor 14 inches (36 cm) thick.

Soldiers use a sight built into the M136 to aim their shots.

The AT4-CS version of the M136 has reduced backblast for use in confined spaces.

BACKBLAST

After the M136 is fired, hot gases come out of the back of the launcher. This creates what is known as backblast, which can injure soldiers in its path. Operators of this launcher must be careful about what is behind them when firing.

ROCKET LAUNCHERS

The SMAW delivers a devastating blast to armored targets.

MK-153 SMAW

The Mk-153 SMAW is the main anti-tank rocket launcher of the Marine Corps. Its name stands for Shoulder-launched Multipurpose Assault Weapon. The SMAW is based on Israel's IMI B-300 rocket launcher, though the design was modified to fit the needs of the Marines. The weapon entered service in 1984.

The SMAW is usually operated by a team of two. The system comes with a spotting rifle. Before the team shoots the missile, they use the rifle to fire special 9-mm tracer rounds that follow the same path the rocket will. Its rounds create a bright flash when they hit the target. Teams use this to make sure that the SMAW will hit the intended target.

USE WITH CAUTION

The launcher is extremely loud, so soldiers nearby need to use hearing protection. Its powerful rockets create a dangerous backblast. Soldiers need to clear the area up to 330 feet (100 m) behind the weapon. This harsh backblast also means that the weapon cannot be fired from inside a building.

Marines reload a SMAW with a blue training rocket in a 2023 exercise.

MINES

M18A1 CLAYMORE

Land mines usually explode upward to injure enemy troops. But the M18A1 Claymore works differently. It explodes to the side. This mine is useful for defending against enemy attacks. It is used by the Army and Marine Corps.

USING THE CLAYMORE

To use the Claymore, a soldier places it on the ground. A sight on top of the mine can be used to aim it in a particular direction. The soldier then moves into a safe position, holding a detonator that is attached to the mine with a wire. When enemy soldiers approach, the soldier sets off the detonator and the mine explodes.

The Claymore contains 700 steel balls and 1.5 pounds (0.7 kg) of an explosive called C4. The explosion launches these balls in a wide pattern in the direction the mine is facing. People within 164 feet (50 m) are likely to be killed. The steel balls remain dangerous to a distance of 820 feet (250 m). The front of the mine features lettering reading "FRONT TOWARD ENEMY." This warning ensures that soldiers always point it in the right direction.

The Claymore has been in widespread use since the 1960s.

Soldiers use a handheld detonator to set off the Claymore at just the right moment.

MORTARS

An M120 mortar round weighs about 30 pounds (14 kg).

M120

Mortars are weapons used to launch explosive rounds at enemy forces. The rounds launch from a tube, traveling in a high arc before striking the enemy. The M120 is one of the US Army's mortars. It is based on the Israeli Soltam K6 mortar. The M120 entered US service in 1991.

The M120 system contains a few parts. In addition to the launching tube, it has a bipod used to hold the tube steady while aiming. It also has a heavy baseplate to give more stability. At 320 pounds (145 kg), it is too heavy for individual soldiers to carry. It must be carried by trailers attached to Humvees. However, it is much easier to transport than full-size artillery.

FIRING

To fire the M120, soldiers drop a 120-mm round down the tube. When it reaches the bottom, it strikes a pin that launches the round out of the tube. The M120 has a maximum range of about 4.5 miles (7.2 km). A trained crew can fire about four rounds per minute. Besides explosive rounds, the M120 can also launch smoke and illumination rounds.

An M120 mortar team fires a round during a training exercise in Japan.

MORTARS

The lightweight M224 mortar is relatively easy for soldiers to move into position and fire.

M224

The M224 is a 60-mm mortar used by the Army and the Marine Corps. The light mortar can be carried by a single soldier. The weapon dates to 1978, when it replaced the World War II–era M2 mortar. The original M224 weighed 46.5 pounds (21 kg). In 2011, the M224A1 model was introduced. It used lighter parts to reduce the weight to 37.5 pounds (17 kg). This makes it even easier to carry across the battlefield.

OPERATION

A crew of three soldiers operates the M224. Like other mortars, the system includes the tube, a bipod, and a baseplate. Soldiers turn screws to adjust the angle and direction of the tube. Working together, a crew can fire 20 rounds per minute. The weapon can also be fired in handheld mode with the smaller M8 baseplate and no bipod. The M224 launches explosive rounds as well as rounds that produce smoke or illuminate nighttime landscapes.

Mortars allow troops to loft rounds over any cover that might be between them and the enemy.

MORTARS

M252

The M252 is an 81-mm mortar in service with the US Army and the Marine Corps. Its weight, firing rate, and range lie between those of the M120 and the M224. The complete unit, including the tube, bipod, baseplate, and sight, weighs 89 pounds (40 kg). A crew of three can keep the M252 firing at a rate of 16 rounds per minute. The mortar has a maximum range of 3.7 miles (6 km). As with other US mortars, soldiers can fire explosive rounds, smoke, and illumination.

Multiple troops work together to rapidly reload and fire the M252.

DEVELOPMENT AND UPGRADES

This medium mortar was developed in partnership with the United Kingdom. The weapon entered service in 1984. A part known as the blast attenuation device (BAD) attaches to the end of the tube. This makes firing the weapon safer for the crew. An upgraded version, the M252A1, was developed later. It is 13 percent lighter than the original model, but it keeps the same range and rate of fire.

A crew working at full speed can go through a large pile of mortar rounds quickly.

85

HOWITZERS

M109

In the US military, cannons larger than mortars are known as howitzers. This term was borrowed from the German language in the mid-1800s. Today, the US has multiple types of howitzers. One of these is the M109. It is used by the Army.

The M109 is a self-propelled howitzer. It runs on tracks, like a tank. This makes the M109 a cross between a weapon and a vehicle. This howitzer fires 155-mm shells to a range of 15 miles (24 km). A total of 39 shells can be stored inside. The M109 also has an M2 machine gun mounted on its roof for close-range defense. It takes a crew of six to operate the M109. The crew consists of a commander, a driver, two gunners, and two loaders.

Soldiers can drive the M109 directly into battle. It does not need to be towed by another vehicle.

HISTORY

The M109 entered service in 1963. It has seen many upgrades in the decades since then. Newer models gave the howitzer more range, added stronger armor, and let it hold more ammunition. The latest model is the M109A7, developed in the 2010s. It adds an automatic loader, meaning the howitzer can have a smaller crew.

An M109 crew prepares to fire during a training exercise in Texas.

HOWITZERS

The M119's wheels allow it to be quickly towed away if needed.

M119

The M119 is a 105-mm howitzer used by the Army. It cannot move under its own power, so it must be brought into action using a vehicle. It is usually towed by a Humvee. The M119 can also be transported by helicopters or dropped from cargo planes using a parachute.

It takes a crew of six soldiers to operate the M119, but in an emergency, four soldiers can keep it firing. A trained crew can shoot the howitzer at a rate of three rounds per minute. These rounds can hit targets up to 10.9 miles (17.5 km) away.

FROM UK TO USA

The M119 is a version of the British L119 howitzer. It entered service in 1989. At first the US military ordered these weapons from the United Kingdom. Later it started building them in the United States. Since the early 1990s, the weapon has had multiple upgrades. These changes improved the aiming system and added rocket-assisted shells that have longer range.

The M119 can be transported by helicopter, providing a lot of flexibility on the battlefield.

The M777 is the latest howitzer in the US arsenal.

M777

The M777 is a 155-mm howitzer. It is towed into place by heavy cargo trucks. It normally takes a crew of eight soldiers to operate the M777. However, five people can use the howitzer in an emergency. A full crew can keep up a firing rate of two rounds per minute.

This howitzer can fire the advanced M982 Excalibur shell. These rounds use the global positioning system (GPS) for extreme accuracy. The M777 can fire them to a range of 25 miles (40 km), with the shells landing within 16 feet (5 m) of the target.

A LIGHTER HOWITZER

Development of the M777 began in the United Kingdom in the 1980s. The goal was to make a lightweight howitzer that would be easier to transport. The design uses titanium, which is lighter than steel. The Army and Marine Corps tested the howitzer in the late 1980s and early 1990s. However, there were problems. The weapon was damaged by firing. After more work, the M777 finally entered service in 2005.

A US Army soldier uses the sights on an M777 as he prepares to shoot during a training exercise.

ROCKET ARTILLERY

A crew launches a rocket at a testing range in Hawaii in 2022.

M142 HIMARS

The name of the M142 HIMARS stands for High Mobility Artillery Rocket System. It is designed to launch rockets at distant enemy targets. Using rockets gives the weapon more range than normal artillery. The M142 entered service with the Army and Marine Corps in 2005.

The M142 has high mobility because it is mounted on large trucks. Soldiers can drive it into position, fire the rockets, and quickly drive to safety. The weapon travels with a resupply vehicle that holds extra rockets. Crews use a small crane to reload the M142.

ROCKET TYPES

The M142 fires multiple kinds of rockets. The M31 carries an explosive warhead up to 44 miles (70 km). The M30A2 is designed to explode 33 feet (10 m) above the ground. It blasts out 160,000 fragments of a hard metal called tungsten. This hits targets over a wide area on the ground. It also has a range of 44 miles (70 km). The MGM-140 ATACMS is used for long-range strikes. Its name stands for Army Tactical Missile System. It can hit targets 186 miles (300 km) away.

Soldiers work together to reload an M142 with a fresh set of six rockets.

ROCKET ARTILLERY

When several M270s fire at the same time, the sky quickly fills with rockets.

M270 MLRS

The name of the M270 MLRS stands for Multiple Launch Rocket System. This rocket artillery weapon is older and larger than the M142 HIMARS. It moves on tracks and is based on the M2 Bradley. The cabin is lightly armored, protecting the crew from gunfire. The M270 entered service with the US Army in 1983.

The M270 fires rockets from large disposable containers. Each container holds six rockets, and the vehicle carries two of them. The system fires the same kinds of rockets as the M142, including the M30A2, M31, and MGM-140 ATACMS. The large ATACMS takes up one whole container. This means an M270 can carry a load of one ATACMS and one pack of six other rockets.

UPDATES

The M270 has gone through upgrades since its introduction. The M270A1 features an improved launch system and can fire newer types of missiles. The M270A2 has a more powerful engine and can be reloaded more quickly.

Each M270 can carry twelve rockets, providing an enormous amount of firepower.

BOMBS

Crews practice loading operations with the AGM-154 so that they can quickly get the weapon into action.

AGM-154 JOINT STANDOFF WEAPON

The term *AGM* stands for Air-to-Ground Missile, but the AGM-154 is actually a bomb. Unlike a missile, it doesn't have any engine of its own. Pilots drop the AGM-154 from an airplane. Then the bomb uses fold-out wings and a GPS guidance system to glide to the target. The bomb can hit targets up to 73 miles (117 km) away from the release point.

The military has multiple types of the AGM-154. The AGM-154A releases 145 small explosives. It is used to destroy light enemy vehicles and missile launchers. The AGM-154C contains a single 500-pound (227 kg) warhead. It is used to attack enemy structures. These weapons are about 13 feet (4 m) long.

TWO BRANCHES

The AGM-154 is used by two branches, the Navy and the Air Force. Many kinds of planes can drop this weapon. This includes the Navy's F/A-18C Hornet and F-35C Lightning II. It also includes the Air Force's F-16 Fighting Falcon and B-2 Spirit.

An F-35C Lightning II drops an AGM-154 during testing.

BOMBS

The head of the GBU-15 contains sensors that help guide the bomb to its target.

GBU-15

The initials *GBU* stand for Guided Bomb Unit. The GBU-15 is dropped from the Air Force's F-15E Strike Eagle jets. It uses small wings and a guidance system to steer toward a target. The weapon has a range of about 17 miles (28 km) from the point where it is dropped.

The GBU-15 has two firing modes. In direct attack, the pilot first locks onto a target. Then the pilot drops the bomb, and the GBU-15 guides itself to the target. In the indirect attack mode, the pilot steers the bomb after dropping it. A camera in the GBU-15 allows the pilot to find and hit a target.

HISTORY

The development of the GBU-15 began in 1974. Its designers applied lessons learned about guided bombs during the Vietnam War. The weapon entered service in 1983. An enhanced version was first used in 2001. It added GPS to help improve targeting.

An F-4 Phantom drops a GBU-15 during training exercises in 1985.

BOMBS

The A-10 Thunderbolt II can carry 16 GBU-39s at once.

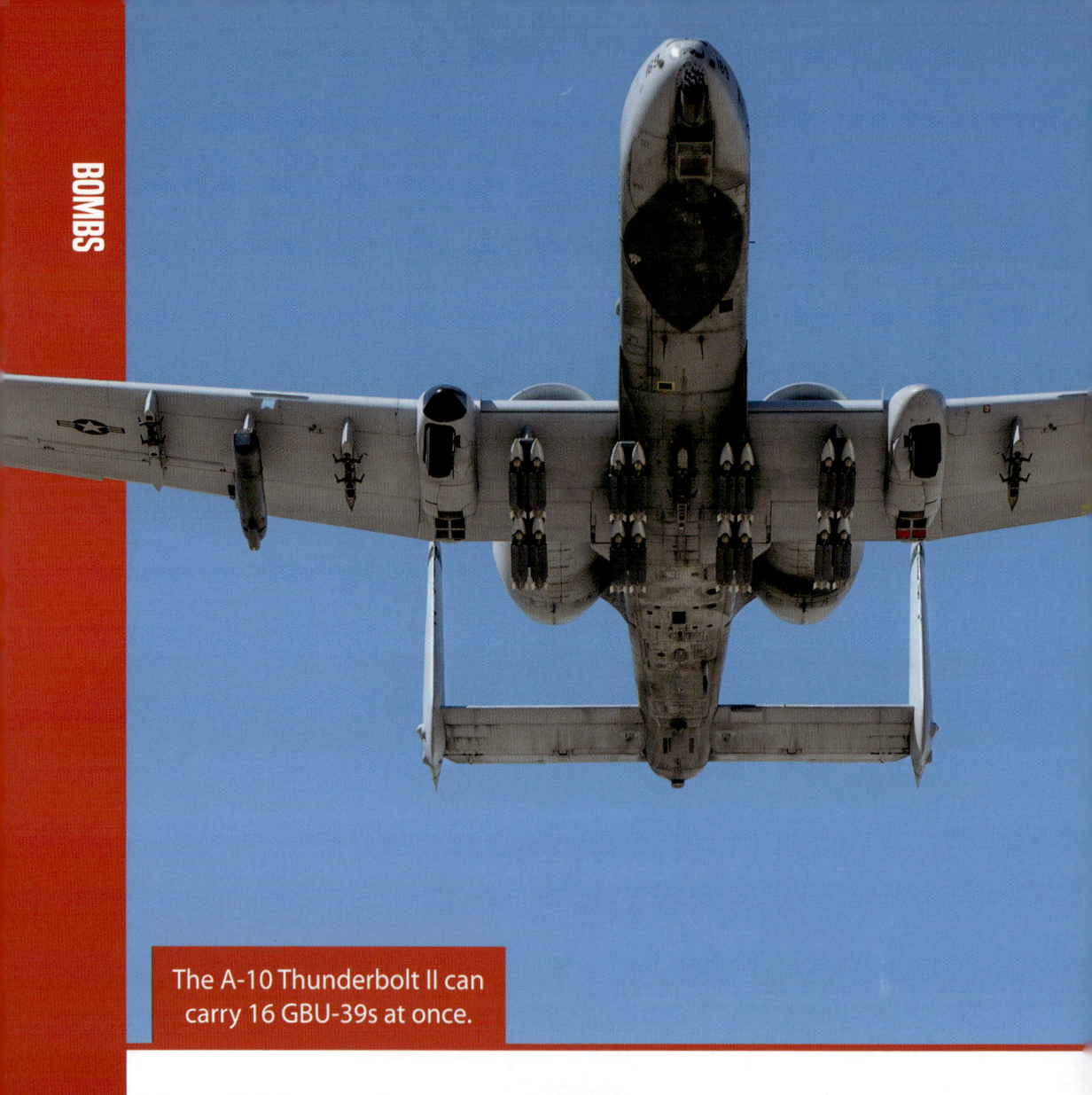

GBU-39 SMALL DIAMETER BOMB

The GBU-39 Small Diameter Bomb is a 250-pound (113 kg) bomb. It works similarly to other guided bombs. After being released from an aircraft, the bomb uses GPS and pop-out wings to guide itself to the target. It can hit targets more than 46 miles (74 km) away from where it is dropped.

Development of the GBU-39 began in 2001. The bomb entered service in 2006. It can be used with the Air Force's F-15E and other jets.

SMALL SIZE

The GBU-39's small size means that airplanes can carry more of them. This lets a pilot strike more targets on a single mission. A plane can release multiple bombs at one time, each one gliding to a different target. The bomb's size has another benefit. It creates a smaller blast than larger bombs. This ensures less damage outside the specific target.

A pilot inspects the GBU-39s on his plane before takeoff.

BOMBS

Two of the people who worked on the MOAB stand near an early test version.

GBU-43/B MOAB

The GBU-43/B MOAB is the most powerful nonnuclear weapon used by the United States. This air-dropped bomb weighs 21,000 pounds (9,525 kg) and is 30 feet (9 m) long. Its name stands for Massive Ordnance Air Blast. However, many people use the nickname Mother of All Bombs.

The MOAB is designed to be dropped from a C-130 Hercules transport aircraft. A parachute pulls the bomb out the back of the airplane. Then fins pop out to steer the bomb to its target. GPS guidance keeps it on track.

HISTORY

The bomb was dropped in combat for the first time in 2017. The Air Force used it against Islamic State in Iraq and Syria (ISIS) fighters in a cave system in Afghanistan. The weapon is useful against enemies in caves. The huge explosion creates a wave of pressure that can injure or kill people deep inside. Officials reported that dozens of ISIS fighters were killed.

The MOAB was developed in 2003 for use in the Iraq War (2003–2011), but it was never used in that conflict.

BOMBS

Workers prepare an early version of the GBU-57 for testing.

GBU-57 MASSIVE ORDNANCE PENETRATOR

The GBU-57 is a type of bomb known as a bunker buster. These bombs are designed to penetrate the surface, digging underground before exploding. They are meant to take out protected enemy bases and bunkers. This huge weapon can be carried by the B-2 stealth bomber. The long-range bomber can carry up to two of these bombs at once.

    The GBU-57 is enormous, weighing about 27,000 pounds (12,250 kg). However, only about 20 percent of that overall weight is explosives. Much of the bomb is made up of hard, dense material. This is what helps it burrow underground before blowing up.

SECRECY

Development of the GBU-57 began in the early 2000s. There are multiple versions of the GBU-57, but the Air Force is secretive about these advanced weapons. Few details are publicly known about how the versions differ. The total number of these bombs in the Air Force inventory is also secret.

A B-52 bomber dropped a GBU-57 over a test range in New Mexico in 2009.

BOMBS

JDAM

The Joint Direct Attack Munition (JDAM) is a kit that turns free-fall bombs into guided bombs. It includes a guidance system and a tail section with fins that can steer the bomb. JDAM can be used with a variety of bombs, with different names based on weight. Attaching it to a 2,000-pound (907 kg) bomb produces the GBU-31. A 1,000-pound (454 kg) bomb creates the GBU-32, and a 500-pound (227 kg) bomb results in the GBU-38.

A JOINT WEAPON

The word *joint* means that multiple military branches use the weapon. The Navy, the Marine Corps, and the Air Force all use JDAMs on their aircraft. Air Force fighters like the F-22 Raptor, as well as the Navy and Marine Corps F/A-18E

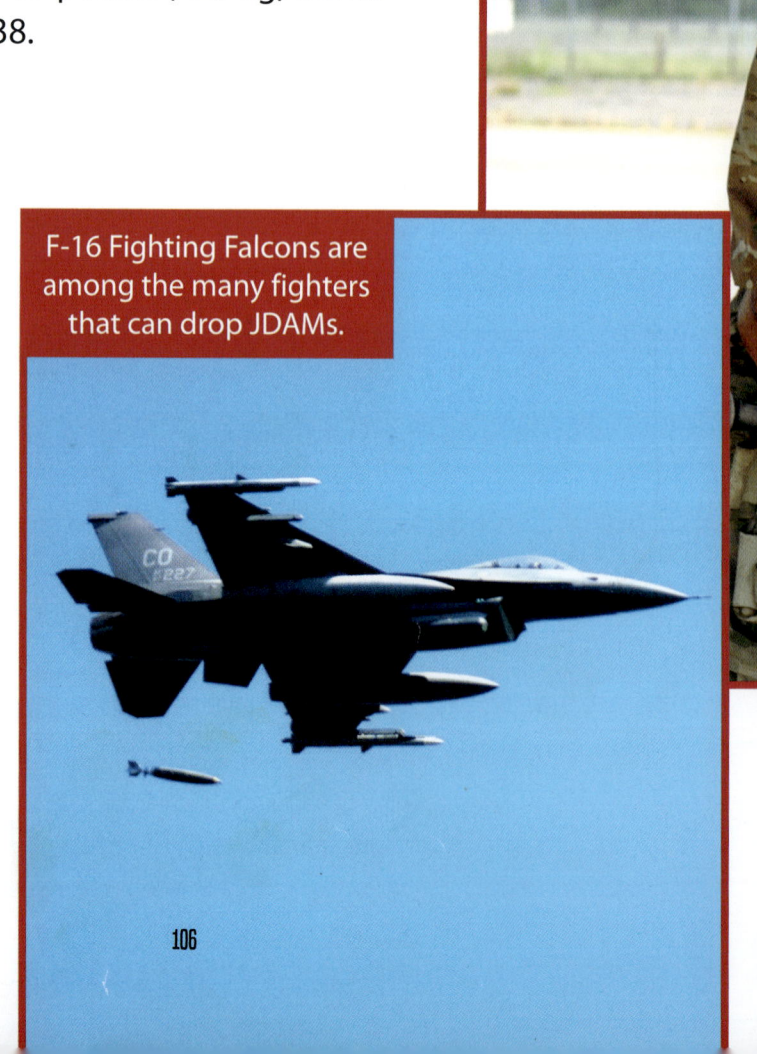

F-16 Fighting Falcons are among the many fighters that can drop JDAMs.

A crew secures a GBU-38 JDAM to an A-10 Thunderbolt II.

Super Hornet, can use the bombs. So can large bombers, such as the Air Force's B-1B Lancer and B-2. The B-2 can release 80 JDAMs on a single pass over a target area.

BOMBS

Air Force personnel add guidance kits, including tail fins, to existing bombs to turn them into guided Paveway bombs.

PAVEWAY

Paveway is a kit that adds guidance and steering equipment to free-fall bombs. It uses laser guidance to hit targets. A laser is aimed at the target. When the bomb falls, it locks onto the laser and steers toward it. The laser may come from the aircraft that drops the bomb, another friendly aircraft, or troops on the ground. Paveway bombs can be dropped from many Air Force jets, including the B-52 Stratofortress, B-2, F-15E, and F-16.

GENERATIONS

There have been three generations of Paveway. Paveway I entered service in the late 1960s. It used fixed wings to steer. Paveway II arrived in the mid-1970s. It used fold-out wings, making it more compact. Its guidance system is more reliable than the earlier version.

Paveway III entered service in the early 1980s. Engineers made improvements that let it fly farther and hit targets more accurately. An upgrade program in 2005 added GPS equipment, making the bombs even more accurate.

Modern versions of the Paveway II are still used on the latest US jets, including the F-35B Lightning II.

AIRCRAFT CANNONS

Removing the GAU-8 from the A-10 and setting it next to a car reveals its impressive size.

GAU-8 AVENGER

The GAU-8 Avenger cannon was built specifically to destroy tanks. This 30-mm cannon was developed in the 1970s. It was designed alongside the Air Force's A-10 Thunderbolt II. This is the only aircraft that uses the GAU-8. The A-10 and its cannon entered service in 1975.

The cannon has seven rotating barrels to give it a high rate of fire. It shoots 3,900 rounds per minute. The plane carries 1,100 rounds of ammunition, meaning a pilot could run through these rounds in less than half a minute. The gun is usually fired in bursts of a second or two.

POWERFUL ROUNDS

The ammunition is a mix of one high-explosive round for every four armor-piercing rounds. The armor-piercing rounds contain a very dense metal called depleted uranium. It is designed to punch through thick tank armor.

The GAU-8 was widely used during the Persian Gulf War (1990–1991). A-10s flew more than 8,000 missions and fired more than 750,000 armor-piercing rounds. The weapon was highly effective against Iraqi tanks.

The GAU-8 is placed slightly off-center on the A-10's nose, with the firing barrel being directly on the center line.

AIRCRAFT CANNONS

Crews must regularly perform maintenance on the M61 to keep it in working order.

M61 VULCAN

The M61 Vulcan is a 20-mm cannon used in many US aircraft. This weapon has six rotating barrels, giving it a high rate of fire. It can shoot 6,000 rounds per minute. The ammunition is a mix of armor-piercing and high-explosive rounds.

    The M61 is mainly used for dogfighting. This is short-range combat where the planes are too close to use missiles. Pilots try to maneuver behind the enemy. Then they shoot a short burst lasting a second or less. The M61 can also be used for attacking ground targets when needed.

HISTORY

Development of this cannon started in 1946, just after World War II. It entered service in the 1950s. The cannon has been used in the F-104 Starfighter of the late 1950s, the F-4 Phantom of the Vietnam War era, and the modern F-15C Eagle and F-16. The Navy's F/A-18 uses it too. The Air Force's F-22 stealth fighter uses a lighter version called the M61A2.

The F-15E's M61 Vulcan cannon is mounted within the aircraft near the base of its right wing.

AIRCRAFT CANNONS

The M134 is commonly mounted on the helicopters of the Army and the Marine Corps.

M134 MINIGUN

In the 1960s, helicopters were becoming an important part of the US military. But crews had trouble defending transport helicopters with existing machine guns. The military requested a new weapon to solve this problem. General Electric made a smaller version of its M61 Vulcan cannon. They called it the M134 minigun. Like the Vulcan, it had six barrels. But instead of the Vulcan's 20-mm rounds, the new weapon fired 7.62-mm rounds. It could fire up to 6,000 rounds per minute.

The M134 has been used on helicopters such as the UH-1 Iroquois and AH-1 Cobra. It is fired by the pilot or copilot. The gun can also be mounted on the side of the helicopter and fired by a crew member. The Navy uses it on boats as well.

DIFFERENT NAMES

The different military branches use different names for the gun. The Army calls it the M134. It is the GAU-2/A in the Air Force and the GAU-17/A in the Navy.

The GAU-17/A provides heavy firepower to the Navy's light river patrol craft.

AIRCRAFT CANNONS

The M230 is mounted below the cockpit of the AH-64 Apache.

M230 CHAIN GUN

The M230 chain gun is a single-barrel cannon used on the Army's AH-64 Apache helicopter. The Apache has a crew of two. A pilot sits in front while a gunner sits in the back. The gunner remotely controls the M230. This 30-mm cannon is useful against enemy troops and lightly armored vehicles.

    The Apache's gunner uses a helmet-mounted display. This technology can be used to point the M230 where the gunner is looking. The cannon can also point straight ahead or be manually controlled. The helicopter generally carries 1,200 rounds of ammunition.

GROUND USE

The M230 has also been adapted for use on ground vehicles. This model is known as the M230LF, meaning *link fed*. Engineers have developed rounds for the M230LF that can be used against drones. The rounds detect when they fly near a drone and explode in midair.

Ground crews check and load the chain gun before flight.

CRUISE MISSILES

A B-52 bomber executes a test launch of an AGM-86B over the desert in Utah.

AGM-86 ALCM

The AGM-86 ALCM is a missile used by the Air Force. *ALCM* stands for Air-Launched Cruise Missile. This missile is dropped from an airplane. Then wings and a tail fold out. The missile propels itself with a jet engine. It moves slower than a missile powered by a rocket engine, but it has a much longer range. The AGM-86 can fly more than 1,500 miles (2,400 km).

This cruise missile was designed for use with the B-52H bomber, which can carry 20 of them. The missile's range allows the bomber to release its weapons while outside the range of enemy defenses. The missiles are small and fly low to the ground, making them hard for enemies to detect.

NUCLEAR AND CONVENTIONAL

The original model of this missile is the AGM-86B, which entered service in 1982. It carries a nuclear warhead. Starting in 1986, some of these missiles had their nuclear warheads replaced with conventional explosives. The new model was called the AGM-86C. This nonnuclear version saw use in the Persian Gulf War. The Air Force retired the AGM-86C in 2019. The AGM-86B remains in service.

The AGM-86's wings and engine allow it to fly extremely long distances.

CRUISE MISSILES

Crews load three AGM-129s onto the wing of a B-52.

AGM-129 ADVANCED CRUISE MISSILE

In 1982, the AGM-86 cruise missile was new. But the Air Force was already worried that new defensive systems would spot it too easily. It began a program to create an improved cruise missile. The result was the AGM-129, which entered service in 1990.

A STEALTHY LONG-RANGE MISSILE

The AGM-129 can carry nuclear warheads. Up to 12 of these missiles can be launched from the B-52H bomber. The aircraft can fire several missiles at once to overwhelm enemy defenses.

Like the AGM-86, this missile has wings and is powered by a jet engine. But it has better range and accuracy compared with the earlier cruise missile. The AGM-129 can travel more than 2,000 miles (3,200 km) from its launching point.

The cruise missile uses stealth technology, making it harder for enemy radar equipment to detect. The shape of the missile helps the weapon deflect radar signals. Special materials in the missile also absorb radar energy.

Stealth features make the AGM-129 more likely to evade enemy defenses and reach its target.

CRUISE MISSILES

AGM-158 JASSM

The AGM-158 JASSM is a stealthy air-launched cruise missile. *JASSM* stands for Joint Air-to-Surface Standoff Missile. It is used by both the Air Force and the Navy. The Air Force can use it on bombers such as the B-2 and fighters such as the F-15E. The Navy's F/A-18E can use it too.

The AGM-158's stealth features help it hide from enemy radar. It uses similar GPS guidance to that used by JDAM bombs. It also stores 3D maps of the target area in the missile's guidance system to help precisely hit the target. The Air Force says the JASSM can land within 10 feet (3 m) of the target point.

EXTENDED RANGE

The AGM-158A entered service in 2009. It has a range of 230 miles (370 km). The upgraded AGM-158B JASSM-ER, or Extended Range, arrived in 2014. This version has a bigger fuel tank and a more efficient jet engine. It can hit enemy targets 620 miles (1,000 km) away.

Even the Air Force's small fighters, such as the F-16, can carry the JASSM.

NON-FLIGHT HARDWARE

Crews practice quickly attaching the JASSM to aircraft.

CRUISE MISSILES

Tomahawk missiles launch vertically from US Navy destroyers.

BGM-109 TOMAHAWK

The BGM-109 Tomahawk is a cruise missile launched from US Navy cruisers, destroyers, and submarines. It entered service in 1983. Early versions carried either nuclear or conventional warheads. However, the nuclear version was later retired.

TOMAHAWK BLOCKS

Models of the Tomahawk are divided into groups called Blocks. The original missiles were Block I. They included a nuclear version and a conventional anti-ship version. Block II was designed to be used against targets on land. One model was used against large targets, such as buildings at naval bases and airfields. Another was for smaller targets with less protection, such as parked aircraft.

Block III Tomahawks added GPS navigation. This upgrade also let the missile wait at a target, circling overhead until the timing was right to attack. Block IV is able to change targets in midair, and it can use cameras to observe the results of previous attacks. The first Block V Tomahawks were delivered to the US Navy in 2021. They feature upgrades to the missile's navigation systems.

Loading a ship with Tomahawk missiles is a major operation.

AIR-TO-AIR MISSILES

The AIM-7 is one of the oldest missiles in the US arsenal.

AIM-7 SPARROW

The AIM-7 Sparrow is a radar-guided air-to-air missile. Fighter jets fire it at enemy aircraft. When the missile is launched, the plane fires radar signals at the enemy plane. The Sparrow's guidance system locks onto these signals, following them to the target. A high-explosive warhead damages or destroys the enemy aircraft.

    The Sparrow contains a guidance system in front, a warhead, a control system, and a rocket motor. Four wings in the middle and four tail fins allow it to steer. The exact range and speed of the missile are classified.

A LONG HISTORY

This missile first entered service in 1956. It was one of the world's first missiles designed to hit enemy aircraft beyond visible range. The Sparrow was used widely during the Vietnam War, but its performance was disappointing. Engineers made many improvements. The latest version is the AIM-7M, which entered service in 1982. It is used by F-15C and F-16 fighter jets.

F-15Cs fire their AIM-7 missiles during an exercise off the coast of Hawaii.

AIR-TO-AIR MISSILES

Marines prepare an AIM-9X for loading aboard a fighter jet.

AIM-9 SIDEWINDER

The AIM-9 Sidewinder is a heat-seeking air-to-air missile. It uses sensors to lock onto the hot exhaust from enemy aircraft. A rocket motor propels it at high speed, and fins steer it toward the target. The missile is used with Air Force, Navy, and Marine fighter jets, including the F-15C, F/A-18E, and F-35. Its exact range and speed are classified.

DEVELOPMENT

The Sidewinder originally entered service in 1956. The early model had several issues. It had a short range, it didn't work well close to the ground, and it couldn't be used at night. Later upgrades fixed these problems.

The AIM-9X entered service in 2003. It has advanced sensors for better guidance. It also has smaller fins, improving its flight performance. The missile works with the Joint Helmet-Mounted Cueing System. This high-tech helmet allows pilots to aim at targets by simply looking at them.

An F-35C carries an AIM-9X under each wing in a 2012 test flight.

AIR-TO-AIR MISSILES

Crews perform safety checks on air-to-air missiles before a fighter jet leaves on a mission.

AIM-120 AMRAAM

Fighter aircraft use the AIM-120 AMRAAM to shoot down enemy planes. *AMRAAM* stands for Advanced Medium-Range Air-to-Air Missile. Before launch, this missile uses its aircraft's radar to lock onto targets. Then the pilot fires the missile. When the AMRAAM gets close to the target, its own radar takes over. The missile finds the target and steers toward it. This self-guidance lets pilots fire multiple missiles at different targets at the same time.

The missile is used by the Air Force, the Navy, and the Marine Corps. US fighters such as the F/A-18C and F-22 are equipped with it. Many US allies also have the AIM-120. The European Eurofighter Typhoon and Swedish JAS 39 Gripen fighter jets both use the missile.

UPGRADES

The AIM-120 entered service in 1991. There have been many upgrades since then. The latest is the AIM-120D3. This model has more efficient computers. This lets the missile's battery last longer, giving it better range. In a 2022 test, an F-15E used the AIM-120D3 to shoot down a remotely controlled aircraft.

An F-15E Strike Eagle from a flight test squadron fires an AMRAAM missile.

SURFACE-TO-AIR MISSILES

The RIM-116 missile leaves its launcher with incredible speed.

RIM-116 ROLLING AIRFRAME MISSILE

The RIM-116 Rolling Airframe Missile is a US Navy weapon originally designed to defend ships against enemy missiles. Later upgrades let it attack enemy planes, helicopters, and surface targets. Its name comes from the way the missile rotates in flight. This helps keep it stable, similar to how a rifle bullet spins. The rolling motion is unique among Navy missiles.

The RIM-116 is used on many kinds of vessels, including aircraft carriers, amphibious assault ships, and littoral combat ships. It is fired from the Mk-49 launcher. The launcher turns and tilts to aim the missiles before firing.

DEVELOPMENT

This missile was developed in partnership with Germany. It was originally based on parts from other weapons. It used the guidance system from the Stinger missile. The warhead and rocket motor came from the Sidewinder missile. The RIM-116 entered service in 1992. Since then, upgrades have improved each of its systems.

The Mk-49 launcher can hold 21 missiles at a time.

SURFACE-TO-AIR MISSILES

Sailors load an Evolved Sea Sparrow Missile launcher aboard an aircraft carrier.

RIM-162 EVOLVED SEA SPARROW MISSILE

The RIM-162 Evolved Sea Sparrow Missile (ESSM) is fired from US Navy ships to shoot down enemy anti-ship missiles. It was developed from the RIM-7 Sea Sparrow, which in turn was based on the AIM-7 Sparrow air-to-air missile. However, the ESSM has so many changes that it is an entirely new weapon. It uses the same launching system as the Sea Sparrow, the Mk-41 Vertical Launching System. This means ships can easily upgrade to the new missile.

The ESSM entered service in 2004. It is used on US destroyers. The missile's large rocket motor gives it a range of more than 31 miles (50 km).

UPGRADES

The original Block 1 model requires the launching ship to fire radar signals at the target to steer the missile. The Block 2 upgrade gives the missile its own radar system. This allows the missile to guide itself, making it easier to hit moving targets.

The ESSM travels at about four times the speed of sound.

AIR-TO-SURFACE MISSILES

ADVANCED PRECISION KILL WEAPON SYSTEM

The Advanced Precision Kill Weapon System (APKWS) gives new abilities to an existing weapon. The Hydra is an unguided rocket 2.75 inches (7 cm) in diameter. Launched from aircraft, its origins date back to the 1940s. In 2012, the APKWS entered service. It combined the Hydra with a new laser guidance system, making it much more accurate.

The APKWS can lock onto targets more than

Each APKWS pod contains several rockets.

Each APKWS rocket is about 6.2 feet (1.9 m) long

3.7 miles (6 km) away. After it is fired, wings pop out to steer the rocket. The launching aircraft points a laser beam at the target, and the APKWS follows the beam. A 10-pound (4.5 kg) explosive warhead destroys the target. The APKWS is cheaper than other guided rocket systems. Its small size also limits collateral damage.

LAUNCHING PLATFORMS

The APKWS was originally used on AH-1W and UH-1Y helicopters flown by the Marine Corps. It later expanded to other helicopters, as well as jets such as the Air Force's A-10 and Marine Corps' AV-8B Harrier II. In 2022, engineers tested a new anti-armor warhead that would let the APKWS destroy tanks.

AIR-TO-SURFACE MISSILES

The AGM-65 provides the Air Force A-10 with accurate long-range firepower.

AGM-65 MAVERICK

The AGM-65 Maverick is an air-to-surface missile. Aircraft fire it at many kinds of ground targets, including armored vehicles, missile launchers, ships, and more. The missile is used on aircraft such as the Navy's F/A-18C and the Air Force's F-16. It entered service in 1972. The Maverick was widely used during the Persian Gulf War to destroy Iraqi tanks.

MAVERICK OPTIONS

There are multiple kinds of Mavericks. All of them use the same rocket motor, but there are three guidance options. The first guidance option uses a camera in the missile. This lets the pilot lock onto a target visually. The second option is similar, but it uses an infrared camera that sees heat. This makes it easier to spot targets at night. The third option uses laser targeting. The missile follows a laser beam aimed by the launching aircraft or friendly forces.

The Maverick also has two warhead options. The first is a 125-pound (57 kg) shaped charge warhead. This is useful against armored targets, such as tanks. The second is a 300-pound (136 kg) penetrator warhead designed to burrow into a target before exploding.

A cover protects the guidance equipment in the Maverick's nose before flight.

AIR-TO-SURFACE MISSILES

AGM-84K SLAM-ER

The AGM-84K SLAM-ER is an air-to-surface missile used against targets on land and at sea. Its name stands for Standoff Land Attack Missile, Expanded Response. *Standoff* means that a weapon can be fired from long range, keeping the attacking aircraft at a safe distance.

Aircraft including the Navy's F/A-18C and P-3 Orion can fire the SLAM-ER. The Air Force's F-15E can also carry this weapon. The missile uses GPS guidance and an infrared sensor to accurately hit targets. It has a range of more than 178 miles (287 km). Pilots can remotely control the missile after launch, redirecting it to a new target if the original one is destroyed.

US Marines load a SLAM-ER onto an F/A-18D Hornet.

An F/A-18C flies with a SLAM-ER, *top*, during testing.

DEVELOPMENT

SLAM-ER entered service in 2000. It is an upgrade to the earlier SLAM missile, featuring improved technology for locating and tracking targets. The SLAM, in turn, was based on the Harpoon anti-ship missile of the 1970s. SLAM-ER upgrades make the missile better at locating and tracking targets.

AIR-TO-SURFACE MISSILES

Multiple HARM missiles can be loaded onto the F-16.

AGM-88 HARM

The AGM-88 HARM is an air-to-surface missile designed to destroy enemy air defense systems. *HARM* stands for High-speed Anti-Radiation Missile. Air defense systems use powerful radars to target aircraft. HARM locks onto these radar signals. The missile's high speed gives enemies little time to notice the missile coming, so they are unable to shut down their radar to confuse the missile's guidance. Destroying air defenses makes the skies safer for friendly aircraft.

HISTORY

HARM missiles entered service in 1984. They were widely used during the Persian Gulf War, with about 2,000 missiles fired at Iraqi air defenses. HARM is used on both Navy and Air Force aircraft, including the F/A-18E and F-16. Later upgrades to the AGM-88F and AGM-88G models improved the missile's guidance system and radar sensors. Further updates will make the missile compatible with the F-35 stealth fighter and the B-21 stealth bomber.

HARM missiles are vital for clearing out enemy air defenses.

AIR-TO-SURFACE MISSILES

AGM-114 HELLFIRE

The AGM-114 Hellfire is a laser-guided air-to-ground missile used for short-range attacks. It can hit targets up to about 6.8 miles (11 km) away. The missile was designed in the 1970s as an anti-tank missile for helicopters, and it entered service in 1986. Later upgrades gave it the ability to target enemy bunkers, radar equipment, and more. It was also adapted for use by MQ-1 Predator drones. The Army, Navy, Marine Corps, and Air Force all use the Hellfire.

HELLFIRE MODELS

There are many models of Hellfire. The AGM-114F added a new warhead to help destroy modern tanks. The AGM-114K improved the targeting system. It allowed the Hellfire to find a target again after losing track of it. The AGM-114L improved the missile's ability to guide itself to a target.

US Navy helicopters can use Hellfire missiles against targets at sea.

A fully loaded MQ-9 Reaper drone can carry eight Hellfire missiles.

    One unique model is known as the AGM-114R9X. Rather than an explosive, it has several steel blades that pop out to destroy a target. This reduces danger to people and structures near the target. The US military has used the R9X to kill terrorist leaders.

AIR-TO-SURFACE MISSILES

AGM-176 GRIFFIN

The AGM-176 Griffin is a small, lightweight air-to-ground missile. It is 43 inches (109 cm) long and carries a 13-pound (6 kg) warhead. It has a range of up to 12.5 miles (20 km). The missile's small size helps to reduce collateral damage. It can use either GPS guidance or laser targeting to hit its targets. The missile entered service in 2001. Its designers reused parts from the Sidewinder and Javelin missiles to keep costs down.

The naval version of the AGM-176 is known as the Griffin C.

Marines prepare a Griffin for firing from a KC-130J aircraft.

FIRING THE GRIFFIN

The Griffin is used on AC-130J gunships. It launches backward from the plane's rear ramp. It can also be fired from the OH-58D Kiowa helicopter and the MQ-9 Reaper drone. The missile is intended for use from airplanes, but the Army has also tested it for ground use. The Navy tested a version of the missile in the 2010s. This anti-ship model was used to target small boats threatening Navy ships.

ANTI-SHIP MISSILES

The Penguin missile is also in service with other countries, including Brazil and Ukraine.

AGM-119 PENGUIN

The AGM-119 Penguin is an anti-ship guided missile. It was developed by Norway in the 1960s, with some funding coming from the United States. The Penguin was used by Norway's navy starting in 1972 and entered US Navy service in 1994. The missile was originally meant for use from ships. However, it later was adapted for use from helicopters such as the SH-60B.

ATTACKING SHIPS

The Penguin uses an infrared sensor to detect heat from enemy ships. Its rocket engine propels it to a range of more than 21 miles (34 km). As the missile nears the target, it begins making random maneuvers. This makes it harder for enemies to shoot it down. The Penguin strikes the enemy ship near the waterline. A delayed fuse means that the 265-pound (120 kg) warhead explodes inside the ship to do maximum damage.

An SH-60B fires a Penguin missile during a 2002 test off the coast of Japan.

ANTI-SHIP MISSILES

The large LRASM can deliver a devastating blow to enemy ships.

AGM-158C LRASM

The name of the AGM-158C LRASM stands for Long Range Anti-Ship Missile. This weapon is based on the AGM-158 JASSM cruise missile. It is upgraded to specifically attack enemy ships.

Development of the missile began in 2009, and the first flight test was in 2013. The weapon entered service in 2018.

The LRASM uses GPS and other sensors to find its targets. Stealth technology helps it hide from enemy detection as it flies toward its target. The missile delivers a 992-pound (450 kg) high-explosive warhead.

WORKING TOGETHER

The Air Force's B-1B can carry up to 24 of these missiles. Smaller jets, such as the F-15E and F-16, can carry just one or two. When multiple LRASMs are launched together, they can share data for a coordinated attack on enemy ships.

In a 2013 flight test, an LRASM without an explosive warhead successfully hit a target ship off the coast of California.

ANTI-SHIP MISSILES

The RGM-84 was the earliest version of the Harpoon missile to enter service.

RGM/AGM/UGM-84 HARPOON

In the 1960s, the Navy recognized the potential importance of anti-ship missiles. It began developing one for US forces. The result was the Harpoon missile. There are three versions of the Harpoon. The RGM-84 launches from surface ships and entered service in 1977. The AGM-84 launches from aircraft and went into use in 1979. The UGM-84 launches from submarines and arrived in 1981.

Harpoon missiles use radar to lock onto an enemy ship. They fly using a jet engine. The ship and submarine versions also have a rocket engine. It boosts them away from the vessel before the jet engine takes over. The missile flies close to the ocean's surface, making it harder for enemy ships to detect.

UPGRADES

New versions over the years added many upgrades. This includes larger fuel tanks and better systems for recognizing targets. Designers borrowed technology from JDAM bombs and SLAM-ER missiles to improve accuracy. The Harpoon's range varies based on the model, with a maximum of about 196 miles (315 km).

A Navy crew loads a UGM-84 missile into a Los Angeles–class attack submarine.

TORPEDOES

Crews use cranes to load and unload torpedoes.

MK-48

The Mk-48 torpedo is designed to sink both submarines and surface ships. This weapon is carried by all US submarines. The 19-foot (5.8 m) torpedo launches from a submarine's torpedo tubes.

The torpedo has two different ways to find its target. One option is to guide it by wire from the submarine. Another is for the torpedo to use its own sensors to find enemy vessels. If it misses, it is able to circle around and try again. The Mk-48 is designed to explode directly beneath ships to cause maximum damage.

HISTORY

The Mk-48 entered service in 1972. In that same year, the Soviet Union began using an advanced new submarine. The Navy decided to upgrade the torpedo to make sure it could defeat the new vessels. It began the Advanced Capability (ADCAP) program, which improved the torpedo's electronics. The Mk-48 ADCAP entered service in 1988.

Sailors inspect and maintain torpedoes while at sea.

TORPEDOES

Mk-50 torpedoes are small and contain advanced technology.

MK-50

The Mk-50 torpedo is used by Navy surface ships and aircraft. At 9.5 feet (2.9 m) long, it is about half the size of the submarine-launched Mk-48 torpedo. It moves quickly and can dive deep to find and destroy submarines. The torpedo carries a 100-pound (45 kg) warhead. It uses sonar to locate its targets.

The torpedo is powered by a stored chemical energy system. Sulfur gas mixes with the metal lithium to produce heat. This heat drives a water jet that propels the torpedo forward.

CHALLENGES

Development of the Mk-50 began in the 1970s. The weapon ended up being complex and expensive, and the designers faced challenges. It finally entered service in 1992, partially replacing the earlier Mk-46 torpedo.

Sailors load a Mk-50 torpedo aboard a guided missile destroyer.

TORPEDOES

Sailors load a Mk-54 torpedo into the weapons bay of a P-8 Poseidon in preparation for a test.

MK-54

The Mk-54 torpedo is fired from surface ships using torpedo tubes or anti-submarine rocket launchers. It is also dropped by anti-submarine aircraft. A GPS-guided parachute system allows planes such as the P-8 Poseidon to deliver the weapon from high above the ocean surface.

The 8.9-foot (2.7 m), 608-pound (276 kg) torpedo entered service in 2004. It uses sonar to detect enemy ships. The weapon's

range is classified. It is designed to work well in shallow waters, sinking enemy submarines that lurk close to the shoreline.

OLD AND NEW

The Navy's Mk-50 torpedo was seen as too expensive. The Mk-54 was developed in response. It combines the advanced sensors of the Mk-50 with the cheaper propulsion system of the Mk-46. The torpedo also shares computer hardware and software with the Mk-48 ADCAP.

Helicopters such as the Navy's MH-60R Sea Hawk can drop the Mk-54 torpedo.

NAVAL MINES

The bomb bay of a B-52 Stratofortress can hold a large number of Quickstrike mines.

QUICKSTRIKE MINE

Naval mines are explosive devices put in a body of water to sink enemy ships or submarines that pass by. These useful weapons prevent enemy vessels from moving freely. The Navy uses a few types of naval mines. One is the Quickstrike family of mines. They are delivered by aircraft into the target area.

Quickstrike mines are mostly adapted versions of normal air-dropped bombs. The Mk-62, Mk-63, and Mk-64 are made from 500-pound (227 kg), 1,000-pound (454 kg), and 2,000-pound (907 kg) bombs. They add a fuse system to normal bombs, designed to set the bomb off if it detects the sound or pressure from passing ships. A fourth version, the Mk-65, is a 2,000-pound (907 kg) mine using a specialized mine casing.

LONG-RANGE MINES

To deliver Quickstrike mines, planes must fly over the spot where they will be dropped. This may put the planes in danger. To solve this problem, the Navy has developed an upgrade called Quickstrike-ER. It combines Quickstrike mines with JDAM-ER guidance kits. This lets planes release mines up to 40 miles (64 km) from the target.

A sailor works on a Mk-65 Quickstrike mine attached to a P-3C Orion.

NAVAL MINES

SUBMARINE-LAUNCHED MOBILE MINE

The Submarine-Launched Mobile Mine (SLMM) gives the Navy a way to secretly place mines in enemy waters. It is especially useful in shallow waters and areas where planes cannot deliver mines. Unlike planes, submarines can place mines even in places with poor weather or surface ice.

The mine is a modified Mk-37 torpedo with a target-detecting device added to it. The submarine fires the SLMM from a torpedo tube. The mine can then travel up to 10 miles (16 km) to its final location. The SLMM weighs about 1,660 pounds (750 kg). About 510 pounds (230 kg) of that is

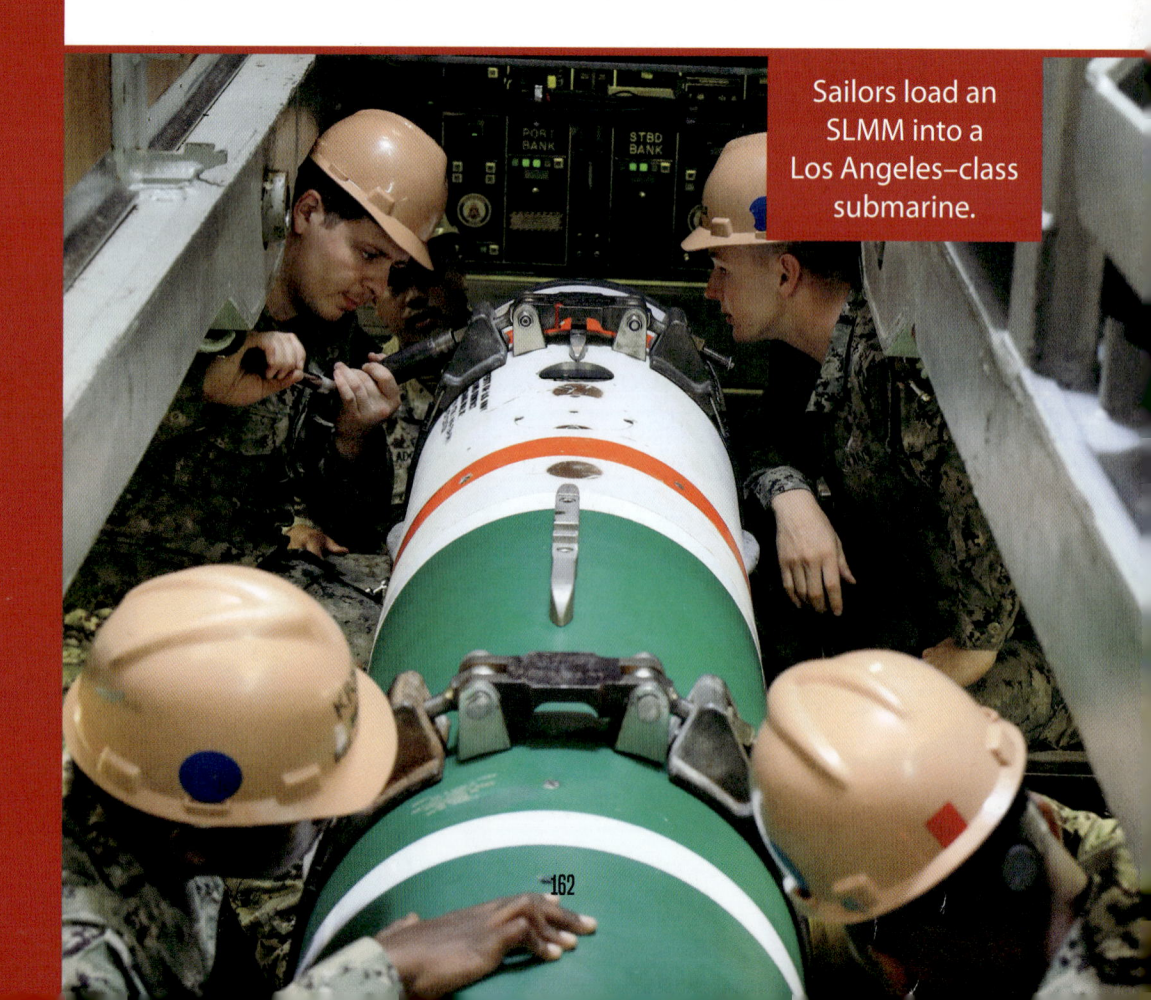

Sailors load an SLMM into a Los Angeles–class submarine.

> The SLMM combines the mobility of a torpedo with the lingering capability of a mine.

its explosive warhead. The mine's trigger can be set off by the magnetism or pressure of a passing ship.

A PLANNED UPGRADE

The Navy once planned to upgrade this weapon by developing mines based on the newer Mk-48 torpedo. However, this idea never moved forward. The SLMM remains the only US naval mine designed to be delivered by submarines.

NAVAL GUNS

A sailor loads ammunition on a Mk-38 gun aboard an amphibious assault ship.

MK-38

The Mk-38 gun fires 25-mm rounds to defend Navy ships against threats from small boats. It is a naval version of the Army's Mk-242 chain gun. When it is installed on the Navy's Mk-88 mount, this gun becomes the Mk-38. The gun has an adjustable rate of fire. It can fire single shots, or it can fire in automatic mode at up to 180 rounds per minute. The weapon entered service in 1986. It is used on most Navy vessels.

A KEY UPGRADE

The original Mk-38 gun was aimed by hand, and it was not stabilized. This made it challenging to hit targets from a moving vessel, especially a small one being tossed by the waves. To solve this problem, the Navy ordered the Mk-38 Mod 2. This upgraded version is stabilized and remotely controlled. The operator uses a screen and a control system to aim and fire.

The Mk-38 Mod 2's remote firing system doubled the accurate range compared with the original gun.

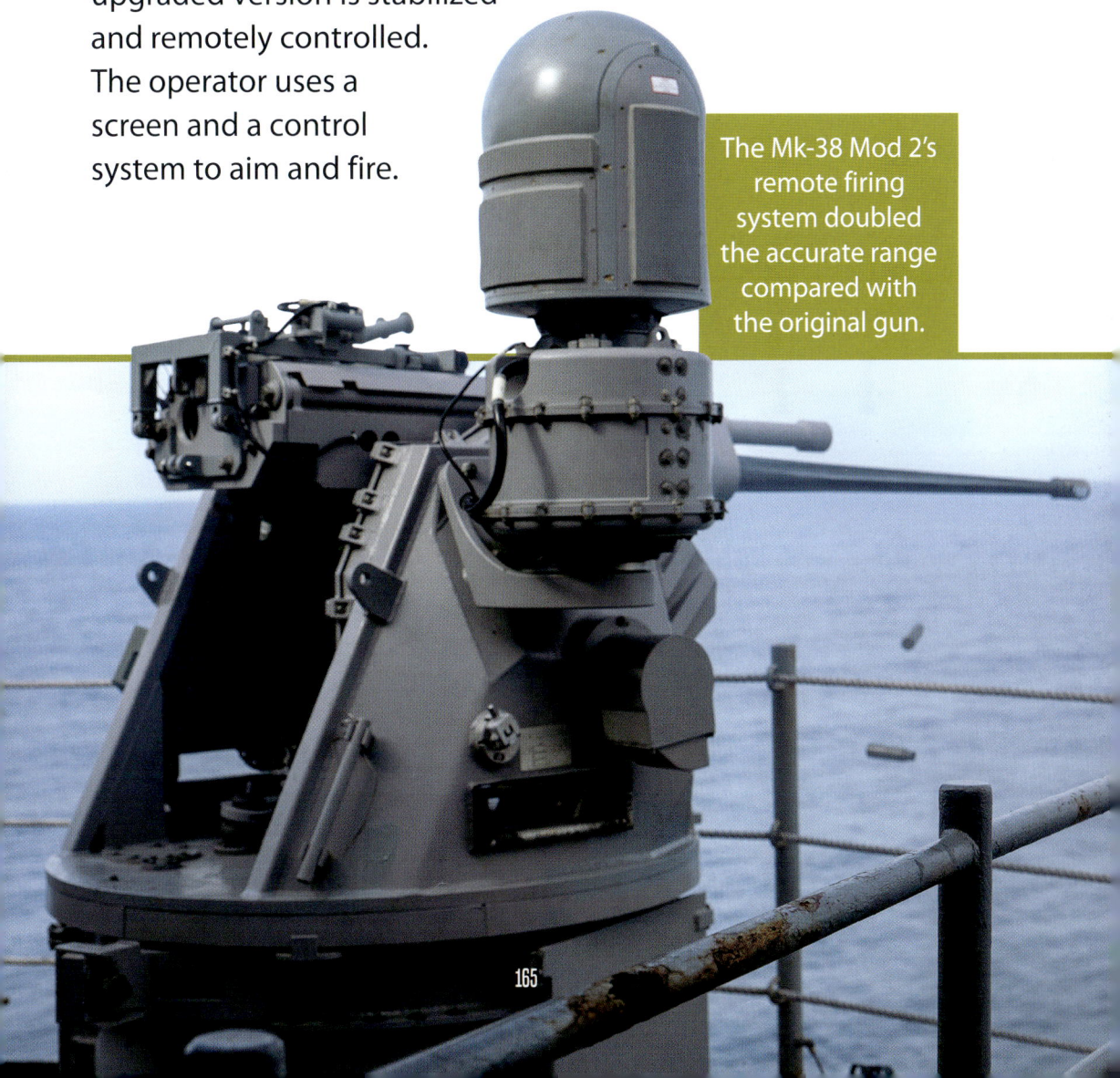

NAVAL GUNS

MK-45

The Mk-45 naval gun fires 5-inch (12.7 cm) shells. The gun has several uses. It can hit enemy ships and aircraft. It can also strike targets on the shore, supporting landings by friendly forces. The gun fires at a rate of up to 20 rounds per minute. It is installed on US cruisers, destroyers, and amphibious assault ships.

MOD 2 AND MOD 4

The original Mk-45 entered service in 1971. Today there are two models of the gun in service. The Mk-45 Mod 2 arrived in the 1980s. The Mk-45 Mod 4 entered service in 2000.

The Mk-45 is the largest naval gun in widespread use with the US Navy.

Crews manage the Mk-45 naval gun from below deck.

These weapons are actually named for their mounting systems. The Mod 2 pairs a Mk-19 gun with the Mk-45 mount. The Mod 4 combines the Mk-36 gun, which has a longer barrel, with the same mount. A longer barrel gives shells more speed as they leave the gun, resulting in better range. The Mod 2 can hit targets 15 miles (24 km) away, while the Mod 4 has a range of 23 miles (37 km).

BALLISTIC MISSILES

LGM-30 MINUTEMAN III

The LGM-30 Minuteman III is an intercontinental ballistic missile (ICBM). These weapons travel extremely long distances. They fly in a huge arc, going high enough to enter space before descending to strike their targets. ICBMs carry nuclear warheads.

Earlier ICBMs used liquid fuel. They had to be filled with fuel at the time of launch, which takes valuable time. Minuteman missiles use solid fuel instead. They are stored with their fuel onboard, so they can be fired on short notice.

The original Minuteman missile entered service in the early 1960s. A new version, Minuteman II, followed in the mid-1960s, and Minuteman III arrived in 1970. Minuteman III has a

The US Air Force tests Minuteman III missiles at Vandenberg Space Force Base in California.

range of more than 6,000 miles (9,660 km). It can deliver W78 or W87 nuclear warheads.

The explosive power, or yield, of nuclear weapons is measured in kilotons (kt). One kiloton equals the power of 1,000 tons (907 metric t) of the explosive material TNT. The W78 has a yield of 335 kt, and the W87 has a yield of 300 kt.

LAUNCHING THE MISSILES

Only the president has the authority to order a nuclear attack. After this order, a crew of two officers in an underground control center would launch a group of missiles. The United States has about 400 Minuteman III missiles in silos in Wyoming, Nebraska, Colorado, Montana, and North Dakota.

Minuteman III crews frequently train on procedures for launching missiles.

BALLISTIC MISSILES

A Trident II launches from an Ohio-class submarine in a 2008 test.

UGM-133 TRIDENT II

The UGM-133 Trident II is a submarine-launched ICBM. It entered service in 1990. Ohio-class submarines are the only US vessels that fire this weapon. Each one can carry up to 20 missiles.

The missiles can be launched while underwater, letting submarines sneak close to enemy shores before firing.

Expanding gas pushes the missile out of the launch tube. Once the missile reaches the surface, its solid-fuel rocket engine fires. It has a range of about 4,600 miles (7,400 km).

MIRV

The Trident II contains multiple independently targetable reentry vehicles (MIRV). This system allows one missile to carry several nuclear warheads that would each strike separate targets. A single Ohio-class submarine can launch more than 150 in all. Each one is able to cause massive destruction. Trident missiles carry W76 or W88 warheads. The W76 has a yield of 100 kt, and the W88 has a yield of 475 kt.

> MIRV systems multiply the destructive power of a single missile, carrying several cone-shaped warheads at once.

NUCLEAR WEAPONS

An F-15E Strike Eagle carries a B61 nuclear bomb, *orange*, in a training exercise.

B61 NUCLEAR BOMB

The B61 nuclear bomb originally entered service in 1968. The latest version is the B61-12, which was first produced in the early 2020s. This update is designed to extend the life of existing B61s. It involves replacing parts such as fuses and batteries.

The B61 is about 12 feet (3.7 m) long. It weighs 825 pounds (374 kg). The bomb can be dropped by a variety of aircraft, including fighters such as the F-16 and bombers such as the B-2. A tail guidance kit can be added to steer the bomb as it falls, giving it more accuracy.

ADJUSTABLE YIELD

The B61-12 has an adjustable yield. It can be set to 0.3 kt, 1.5 kt, 10 kt, or 50 kt. Setting a smaller yield could help reduce damage around the targeted area.

An access panel allows ground crews to arm the bomb and select its yield.

NUCLEAR WEAPONS

B83 NUCLEAR BOMB

The B83 nuclear bomb is the most powerful weapon in the US arsenal. It has an adjustable yield, with a maximum of 1,200 kt. This enormous explosive force is enough to destroy an entire city. The bomb is designed to attack tough targets, such as underground enemy ICBM silos.

The B83 was developed in the 1970s and entered service in 1983. The bomb is 12 feet (3.7 m) long. It weighs about 2,400 pounds (1,090 kg). It can be carried by many US aircraft,

Since 1945, the only nuclear weapons detonated have been those used for testing.

Air Force crews use training models of the B83 to practice loading it on airplanes for transport.

including the B-1B and F/A-18C. The bomb is not guided as it falls, but a parachute can slow its drop to give the aircraft time to escape the blast.

FALLOUT

Using the B83 against silos and similar targets would involve exploding it close to the ground. This would create massive amounts of radioactive debris, or fallout. Dangerous fallout would extend over a wide area, sickening or killing many people. For this reason, some critics say the B83 should be retired.

MISSILE DEFENSE

A destroyer equipped with the Aegis BMD system launches an SM-3 missile at a test target.

AEGIS BALLISTIC MISSILE DEFENSE

The Aegis Ballistic Missile Defense (BMD) system is designed to shoot down enemy missiles before they can reach the United States. To detect missiles, the system uses the advanced SPY-1 radar. SPY-1 uses multiple sensors to give a 360-degree view of its surroundings.

Aegis BMD then uses SM-3 or SM-6 missiles to intercept enemy weapons. SM-3 missiles reach extremely high altitudes, destroying enemy ICBMs while they are in space. SM-6 missiles work at lower altitudes, hitting enemy missiles later in flight. Together, these missiles provide wide-ranging protection.

WHERE AEGIS BMD IS USED

Aegis BMD is installed on more than 40 Ticonderoga-class cruisers and Arleigh Burke–class destroyers. Some Japanese ships also use Aegis BMD. A system known as Aegis Ashore uses the same equipment for missile defense on land. An Aegis Ashore installation is located at a military base in Romania. It is operated jointly by Romanian and US troops.

The Aegis Ashore facility in Romania is designed to protect Europe from missile attacks.

MISSILE DEFENSE

Ground-based interceptor missiles are tested at Vandenberg Space Force Base.

GROUND-BASED MIDCOURSE DEFENSE

The Ground-based Midcourse Defense (GMD) system is designed to protect the United States from an ICBM attack. The system includes a variety of sensors located around the globe. When these sensors detect a threatening launch, a missile called a ground-based interceptor (GBI) is fired. GBIs are based in Alaska and California. There are about 40 of them in all. They are able to defend against a limited attack, rather than a large wave of enemy missiles.

INTERCEPTING A MISSILE

The large, powerful GBIs have a broad enough range to protect all 50 US states. A GBI flies into space, maneuvering into the path of an enemy ICBM as it travels toward a US target. Then the GBI releases an Exoatmospheric Kill Vehicle (EKV). This device uses advanced sensors and its own motors to steer into the enemy missile. It does not contain an explosive warhead. Instead, it destroys the missile by simply colliding with it at high speed.

Crews at Fort Greely in Alaska operate the GMD system based there.

MISSILE DEFENSE

MIM-104 PATRIOT

The MIM-104 Patriot is a US Army system that launches missiles to shoot down enemy aircraft and missiles. A single Patriot system, known as a battery, includes many parts. Besides the missile launchers themselves, it has radar gear, a control system, a power generator, and support vehicles. Operating a Patriot battery requires about 90 soldiers.

The Patriot entered service in 1982. It was originally meant to shoot down only aircraft. In the late 1980s, upgrades allowed it to intercept missiles as well. Many more upgrades in the following years improved its power and accuracy.

A powerful radar is a key part of the Patriot air defense system.

Test launches of the Patriot help the Army evaluate and improve the system.

MISSILE TYPES

Patriot launchers use two main missile types. The first is the PAC-2. These missiles contain an explosive warhead to destroy targets. The second type is the PAC-3. These missiles have no warhead. Instead, they simply collide with targets to destroy them. The lack of a warhead means PAC-3 missiles are lightweight, weighing only a third as much as PAC-2. This gives PAC-3 a greater range.

MISSILE DEFENSE

MK-15 PHALANX CIWS

The Mk-15 Phalanx CIWS is the last line of defense for US Navy ships. *CIWS* stands for Close-In Weapon System. It is used to destroy aircraft, missiles, or small boats. The system is an M61A1 20-mm gun mounted on equipment that aims and fires at targets. It entered service in 1980. The Phalanx is used on most US Navy ships.

DEFENDING THE SHIP

The Phalanx is used when threats slip past a ship's outer defenses. When they

> Some sailors have a nickname for the Phalanx CIWS: R2-D2. Its radar dome looks a bit like the robot character from *Star Wars*.

> Sailors reload a Phalanx CIWS aboard an amphibious dock landing ship.

get close, there is no time for crews to aim and fire at them. The Phalanx uses its own radar system to automatically detect, track, and shoot at incoming threats. It fires at an extremely fast rate, up to 4,500 rounds per minute, to improve the chances of destroying the target. The Phalanx has a range of about 3,940 yards (3,600 m).

MISSILE DEFENSE

NASAMS is manufactured at a facility in Kongsberg, Norway.

NASAMS

NASAMS is the National Advanced Surface-to-Air Missile System. This missile defense system was developed by a US and a Norwegian company working together. It is used to shoot down both aircraft and missiles.

There are three major parts to NASAMS. The first is a radar system. It can spot targets up to 47 miles (75 km) away. It is able to identify whether possible targets are friendly or not. The next part is a control station. Soldiers use screens and control panels to operate the system. The final part is the missile launchers themselves. NASAMS fires AIM-120 AMRAAM missiles. These are also commonly used by fighter aircraft, but no changes are required to use them in the ground-based launcher. The missiles have a range of about 25 miles (40 km).

DEFENDING THE CAPITAL

NASAMS originally entered service with the Norwegian military in 1994. It is used to protect military targets as well as civilian areas. In the United States, NASAMS has helped defend Washington, DC, since 2005.

A Norwegian launcher fires a NASAMS missile during a joint exercise with US forces in 2023.

MISSILE DEFENSE

THAAD

THAAD is the Terminal High-Altitude Area Defense system. It is designed to shoot down enemy missiles during their terminal, or final, part of flight. A THAAD battery includes a powerful radar and a set of six launchers. Each launcher contains eight missiles. The missiles have no warheads. Instead, they destroy targets by simply colliding with them.

THAAD has a range of up to 124 miles (200 km). It fills a middle role in US missile defense. It covers a broader area than Patriot missiles but a smaller range than the Aegis BMD and GMD systems.

THAAD launchers are mounted on large trucks, making them easy to transport.

The military carries out THAAD tests on the remote Pacific island of Kwajalein.

DEVELOPMENT AND DEPLOYMENT

This system was developed in the 1990s. The first six tests, held between 1995 and 1999, failed. But development continued. An upgraded version went through testing in the 2000s and 2010s with much more success. THAAD batteries are now active in Guam and South Korea to defend against possible missile threats from North Korea.

GLOSSARY

bayonet
A bladed weapon attached to a gun.

bipod
A two-legged device that supports a weapon to increase stability.

classified
Kept secret for reasons of military security.

collateral damage
The destruction of something other than the intended target.

conventional
Nonnuclear.

fragmentation
Breaking into many small pieces.

global positioning system (GPS)
Technology that uses satellites in orbit and receivers on the ground to help pinpoint locations on Earth.

iron sights
Parts mounted on top of a gun to help the shooter aim.

littoral
Operating in the ocean near the shoreline.

magazine
A part that holds ammunition and is attached to a gun.

radar
A system that locates distant objects by sending out radio waves and detecting the reflected energy.

recoil
The kickback from a gun while firing.

semiautomatic
Firing once and reloading each time the trigger is pulled.

sheath
A cover used to hold a knife.

sniper
A soldier whose job is to hit enemies at long range, usually using a rifle with a scope.

suppressor
A device that reduces the sound of a gunshot.

tracer
A type of ammunition that leaves a bright trail, making it easier to see where rounds are going.

tripod
A three-legged device that supports a weapon to increase stability.

TO LEARN MORE

FURTHER READINGS

Henzel, Cynthia Kennedy. *US Army*. Abdo, 2021.

Huddleston, Emma. *Life in the US Navy*. ReferencePoint, 2021.

Ringstad, Arnold. *The Military Vehicles Encyclopedia*. Abdo, 2024.

ONLINE RESOURCES

Booklinks
NONFICTION NETWORK
FREE! ONLINE NONFICTION RESOURCES

To learn more about US military weapons, please visit **abdobooklinks.com** or scan this QR code. These links are routinely monitored and updated to provide the most current information available.

INDEX

Afghanistan War, 11, 32, 49, 103
aircraft cannons, 4, 110–117
air-to-air missiles, 4, 126–131, 134
air-to-surface missiles, 4, 122, 136–147
amphibious assault ships, 132, 166
anti-ship missiles, 4, 124, 134, 141, 147, 148–153
assault rifles, 4, 14–23, 50, 61, 63

backblast, 75, 77
ballistic missiles, 4, 5, 168–171, 176
bayonets, 4, 6–9, 18
bombers, 4, 104, 107, 118–119, 121, 122, 143, 172
bombs, 4, 96–109, 122, 153, 161, 172–175
Browning, John, 45

carbines, 6, 16, 19, 20, 22, 24–25, 34, 42, 61, 63
chain guns, 4, 116–117, 164–165
cruise missiles, 4, 118–125, 150
cruisers, 124, 166, 177

destroyers, 124, 134, 166, 177
drones, 71, 117, 144, 147

fallout, 175
fighter jets, 4, 106, 113, 122, 126–131, 143, 172, 185

global positioning system (GPS), 90, 96, 99, 100, 102, 109, 122, 124, 140, 146, 151, 158

grenade launchers, 4, 18, 60–65
grenades, 4, 52–59, 60–61, 62, 64
gun accessories, 11, 13, 16, 21, 27, 33, 37

handguns, 4, 10–13
helicopters, 4, 52, 65, 66, 70, 88, 114–115, 116, 132, 137, 144, 147, 148
helmet-mounted displays, 116, 129
howitzers, 4, 86–91

infantry fighting vehicles, 47
Iraq War, 11, 49

jet engines, 118, 121, 122, 152–153

Korean War, 60

laser guidance, 108, 136–137, 139, 144, 146
littoral combat ships, 132

machine guns, 4, 21, 44–51, 86, 114
mines, 4, 78–79
missile defense, 176–187
missile silos, 169, 174–175
mortars, 80–85, 86

naval guns, 4, 164–167
naval mines, 160–163
nuclear weapons, 5, 119, 121, 124, 168–175

Persian Gulf War, 111, 119, 138, 143

radar, 5, 121, 122, 126, 130, 135, 142–143, 144, 152, 176, 180, 183, 184, 186
rocket artillery, 92–95
rocket launchers, 4, 66–77

scopes, 17, 18, 21, 22, 26, 30–31, 33
shaped charges, 74, 139
shotguns, 4, 34–39
smoke grenades, 52–53, 56–57, 61
special forces, 14, 25, 41, 43
stealth, 104, 113, 121, 122, 143, 151
submachine guns, 40–43
submarines, 4, 124, 152, 154–156, 158–159, 161, 162–163, 170–171
surface-to-air missiles, 132–135, 184–185

tanks, 47, 66, 68–69, 73, 76, 86, 110–111, 137, 138–139, 144
torpedoes, 4, 154–159, 162–163

Vietnam War, 18, 60, 65, 99, 113, 127

World War II, 5, 60, 82, 113

PHOTO CREDITS

Cover Photos: Wikimedia Commons, front (M9 Bayonet); US Marine Corps/DVIDS, front (M9 pistol, soldier holding M16A4, M67 grenade); Idaho Army National Guard/DVIDS, front (M107 rifle); US Army, front (soldier holding M26); US Army/DVIDS, front (MP5K gun); US Department of Defense, front (XM250 gun), back (left); US Air Force/DVIDS, front (M83 grenades, M3 rocket launcher, F-15E Strike Eagle, smoke); US Navy, front (F/A-18C Hornet), back (right); US Air Force, front (claymore)

Interior Photos: Idaho Army National Guard/DVIDS, 1, 28–29 (bottom), 30–31 (left), 46; Shutterstock Images, 3; US Air Force/DVIDS, 5, 26, 27, 43, 56, 61, 72–73, 73, 96–97 (top), 100–101, 101, 108, 118, 118–119, 122, 123, 130–131, 138, 142–143, 145, 160, 169, 172–173, 178; Wikimedia Commons, 6, 148; US Marine Corps, 7, 8–9, 9; US Navy/DVIDS, 10, 16–17, 37, 96–97 (bottom), 112, 114–115, 124, 125, 132, 133, 134, 134–135, 139, 143, 144, 152, 153, 154, 155, 158, 158–159, 162, 164, 165, 166, 166–167, 170, 176, 177, 182, 183; US Marine Corps/DVIDS, 11, 18, 18–19, 20, 21, 28–29 (top), 36, 38, 39, 42, 49, 54, 55, 58, 59, 64, 64–65, 68, 69, 70, 76, 77, 84–85 (top), 108–109, 114, 128, 130, 136, 137, 140, 147; US Army/DVIDS, 12, 13, 14, 15, 17, 23, 24–25, 30–31 (right), 32–33, 35, 40, 41, 44, 46–47, 48, 62, 66, 67, 71, 75, 78, 79, 80, 81, 82, 88–89, 89, 90, 91, 92, 93, 94, 113, 116–117, 117, 180, 181, 186–187 (bottom); US Department of Defense, 22–23, 50, 103, 104; US Air National Guard/DVIDS, 24, 74, 106–107, 111; US Army National Guard/DVIDS, 33, 63, 84–85 (bottom), 86; US Army, 34, 44–45; Nathan Laine/Bloomberg/Getty Images, 50–51; US Army Cadet Command/DVIDS, 52–53; Sal (Tony) Lopez/DVIDS, 53; US Army National Guard, 56–57; Melissa Buckley/DVIDS, 60; Markeith Horace/DVIDS, 82–83; North Carolina National Guard/DVIDS, 87; Bryan Araujo/DVIDS, 95; National Archives, 98, 161; US Air Force, 98–99, 102, 104–105, 110, 120–121 (top), 120–121 (bottom), 126, 127, 168, 173, 175; Kevin Clark/DVIDS, 106; US Navy, 129, 141, 146, 149, 150, 156, 157, 163; DARPA, 150–151; Millard H. Sharp/Science Source, 171; PhotoQuest/Archive Photos/Getty Images, 174; Alaska Army National Guard/DVIDS, 178–179; Petter Berntsen/AFP/Getty Images, 184; Royal Norwegian Navy/DVIDS, 185; US Missile Defense Agency/DVIDS, 186–187 (top)

ABDOBOOKS.COM

Published by Abdo Reference, a division of ABDO, PO Box 398166, Minneapolis, Minnesota 55439. Copyright © 2024 by Abdo Consulting Group, Inc. International copyrights reserved in all countries. No part of this book may be reproduced in any form without written permission from the publisher. Encyclopedias™ is a trademark and logo of Abdo Reference.

102023
012024

THIS BOOK CONTAINS RECYCLED MATERIALS

Editor: Walt K. Moon
Series Designer: Colleen McLaren
Production Designer: Karli Kruse

LIBRARY OF CONGRESS CONTROL NUMBER: 2023939639

PUBLISHER'S CATALOGING-IN-PUBLICATION DATA

Names: Ringstad, Arnold, author.
Title: The military weapons encyclopedia / by Arnold Ringstad
Description: Minneapolis, Minnesota: Abdo Reference, 2024 | Series: US military encyclopedias | Includes online resources and index.
Identifiers: ISBN 9781098293055 (lib. bdg.) | ISBN 9798384910992 (ebook)
Subjects: LCSH: Military weapons--Juvenile literature. | Armaments--Juvenile literature. | Military history--Juvenile literature. | United States--Armed Forces--History--Juvenile literature. | Encyclopedias and dictionaries--Juvenile literature.
Classification: DDC 355.82--dc23